AT THE BASE OF THE GIANT'S THROAT

AT THE
BASE
OF THE
GIANT'S
THROAT

The Past and Future of America's Great Dams

ANTHONY R. PALUMBI

Potomac Books
An imprint of the University of Nebraska Press

All rights reserved. Potomac Books is an imprint of the
University of Nebraska Press.
Manufactured in the United States of America.

Library of Congress Cataloging-in-Publication Data
Names: Palumbi, Anthony R., 1984–, author.
Title: At the base of the giant's throat: the past and
future of America's great dams / Anthony R. Palumbi.
Description: Lincoln: Potomac Books, an imprint of
the University of Nebraska Press, 2023 | Includes
bibliographical references and index.
Identifiers: LCCN 2022043565
ISBN 9781640124936 (hardback)
ISBN 9781640125872 (epub)
ISBN 9781640125889 (pdf)
Subjects: LCSH: Dams—United States—History. | BISAC:
HISTORY / United States / General | TECHNOLOGY &
ENGINEERING / Environmental / Water Supply
Classification: LCC TC556 .P35 2023 |
DDC 627/.80973—dc23/eng/20221205
LC record available at https://lccn.loc.gov/2022043565

Set in Vesper by L. Auten.

For my grandfather, Robert Palumbi (1927–2020):
U.S. Army veteran, schoolteacher, and disciple of history

CONTENTS

ACKNOWLEDGMENTS

Nothing I write would ever see the light of day if not for the love and support I enjoy every day from my family. Thank you to my immediate family, Steve and Mary, Lauren and Kevin; my in-laws, Judy and Beto, Robert and Diana; and to the rest of my gorgeous and sprawling kin, I am indebted to you as well. Thanks to Paul Roberts for your care and expert eye. Thanks to my agent, Dawn Frederick, who believed in my work before almost anyone. Thanks to the editor of this book, Tom Swanson, and the dedicated professionals at Potomac Books and the University of Nebraska Press for their work in producing books that matter. Thanks to the Sacramento Public Library system, a vital resource in the research for this book and a public institution in the finest American tradition.

Last, with love and devotion I thank my wife, Liz Palumbi. Anything I could say, you already know. Thank you for your relentless support, for everything you've done and will do.

AT THE BASE OF THE GIANT'S THROAT

1

RIVERS WILD

Every river is a story with a beginning, a middle, and an end, a procession of settings and scenes, characters, and conflicts driven by the action of water. A river is not the water it contains; the water, only a guest, moves from cloud to earth to ocean according to the season. Nor is a river the ribbon of land over which water moves, for the bed is constantly remade and a single day of rain may lead the water to a new course entirely. A river is not the plants and animals along its banks, or the people who drop plumb lines into murky fishing holes, or the memories their children compile playing in the pungent black muck among the cattails.

A river is every stream that fed it, every dip and turn along its course, every field irrigated for a hundred miles around. It is the power plants drawing cold water through cavernous penstocks, the millions of people turning their faucets to receive instant delivery of life's most precious resource. A river is the safety and security we experience every day, the guarantees that let us feel comfortable planning out our lives. It is everything we see and so much more that we cannot. Every river is a story.

By official reckoning, the American story starts in the early 1600s with the Jamestown colony. Of course the continent had been thoroughly inhabited for at least ten thousand years, supporting uncounted millions of Indigenous people in every imaginable environment from the parched landscapes of Arizona to the endless plains of the Dakotas and the dense marshes of

Massachusetts. Indigenous peoples lived in every state in great numbers, for the most part enjoying lives of good health, comfort, and plenty compared to European peasants. The first men who ambled off their ships onto North American soil found their counterparts strikingly tall and well muscled, their faces unblemished by the smallpox scars that cratered nearly every visage in Europe, their clothing and hair maddeningly cool in its color and style. Indigenous Americans were content to trade with these new visitors and sought alliances as leverage against their rivals. Hundreds of tribes were laid out over the land, trading and warring and merging with the winds of fate and politics.

Like all human civilizations, the first Americans gathered around sources of fresh water. Some built permanent settlements while others migrated between summer and winter grounds, following herd animals and fleeing flood plains. The paths they wore through forests and over mountain passes were followed centuries later by European colonists who thought them laid down by Providence. The false claim that Indigenous peoples had built nothing in their time on the continent became an enduring rationale for European land theft. Right-wing intellectual icon Ayn Rand presented this idea in refreshingly unvarnished terms: "The Indians did not have any property rights—they didn't have the concept of property." Thus, she argued, "they didn't have any rights to the land" and "any white person who brings the elements of civilization had the right to take over this continent."

In fact, the conquest of North America and the near total annihilation of its Indigenous population might be the greatest collective crime in world history. But as Charles Mann describes in his vital work 1491, much of the genocide happened out of European sight. European-imported diseases, smallpox especially, ravaged the native population with terrifying speed. The diseases moved so quickly they actually ran ahead of the colonizers, first attacking tribes who interacted directly with Europeans before spreading to neighbors and trading partners who had never even met these pale people from across the ocean. All of them died, crushing entire cultures in the space of a generation. By the time Europeans moved into their homelands, the Natives had already been destroyed. What few they encountered were scattered, impoverished; they were the traumatized survivors of an apocalypse they had no way to understand.

Long before cities of stone and iron rose on the Atlantic coast, trappers and explorers were greedily plumbing the inland reaches. They encountered endless forests, mountains to rival the tallest Alps, and rivers grander than any in Europe. Examine a map of European river basins and you will see a crowded quilt: small rivers with short runs, crowded tightly like sunbathing tourists along the coasts. North American drainage basins are enormous by comparison. Of the world's fifteen largest rivers by average discharge, eight are located in South America and four in Asia. One is in Africa, and North America boasts two, but even the biggest waterway in Europe (the Volga River) cannot crack the top thirty.

The great rivers of North America carve the continent into geographic cantons, to a larger degree and with greater ecological consequences than any mountain range. The Mississippi River basin, to pick the most impressive example, drains more than 40 percent of the United States and discharges it all into the Gulf of Mexico through a single titanic mouth near New Orleans.[1] Mountains trap and shape precipitation, but rivers decide what's done with it. The new civilization rising from the ruins of the great Indigenous nations was a patchwork of European settlements and Indigenous territory, crackling with tension as settler populations sought natural resources, and the natives sought to resist them. The lands of the major river basins were the most precious cropland available, so the United States sunk its earliest foundation in their rich black soil.

St. Lawrence Basin

Water is a remarkable little compound: hydrogen and oxygen, two of the universe's most common elements, fixed in a vital partnership. Oxygen is a corrosive material, overburdened with electrons and negatively charged. It was the natural world's first poison, a waste product that slowly permeated the Earth's atmosphere until life adapted to not just tolerate oxygen but use it as fuel. Two positively charged hydrogen atoms mellow it out—they neutralize oxygen's propensity for chaos, yielding H2O. The three atoms cling together in a shallow V shape, the arms of the V represented by hydrogen with oxygen at their junction. Since both positive charges are on the same side of the molecule, its electric charge is fundamentally unbalanced. Chemists call this "polarity." Many foreign substances dissolve in water because its unbalanced polarity attracts both positive and negative atoms, offering them alternatives to their present bonds, tempting them gradually away from the herd. Salt is the classic example: positively charged sodium and negatively charged chlorine snap together at clean right angles to create cube-shaped crystals. Introduce some water and these formerly staid partners each find something to like, the sodium chasing oxygen while chlorine pursues hydrogen. Heat the water up and these interactions happen even faster—the salt dissolves almost immediately. Water is so cleanly unbalanced that enough of it can dissolve almost any naturally occurring compound, while the chemical bonds of the water itself stay nearly unbreakable.

Another polar property of water is how, at low temperatures, its hydrogen atoms become more strongly attracted to other hydrogen atoms. Liquid water allows molecules to form, break, and reform bonds in haphazard fashion as they slosh around. But if the water is cooled, they rapidly become monogamous. Every water molecule seeks two ironclad bonds for its two hydrogen atoms, eagerly bonding with neighboring water molecules, each becoming a vertex on the most efficient shape geometry will allow their interlocking v-shaped bonds to form: a hexagon. That six-sided structure occupies more space than the six molecules would occupy if dumped together in a liquid state. Most compounds become progressively denser as they transition from gas to liquid to solid, but ice is actually less dense than water. When water freezes, it *expands*.

That expansion can be quite destructive. Water seeps into fissures, cracks, and pores in solid materials and, when it freezes, it acts as an expanding wedge, forcing them wider. Anyone who lives in a wintery climate notices the fresh potholes yawning in their local roads each spring—water has seeped into the roadways' imperfections and fractured them, levering up chunks of asphalt and scattering them across the frozen expanse. If you understand winter potholes, you understand the formation of the Great Lakes.

About 2.6 million years ago, the planet's temperature took a prolonged nosedive. The resulting "ice age" froze large portions of North America under a massive complex of glaciers known today as the Laurentide Ice Sheet. The expansionary properties of ice, combined with the glaciers' seismic weight and constant downhill movement, gouged deep holes and fissures into the planet's surface. When the planet warmed again, the Laurentide retreated, the ice melted away, and fresh water pooled in the glacial scars that remained. Thus the Great Lakes came to be. The glaciers had been so heavy that they compressed the entire region below sea level, but once relieved of the ice's awful weight the land began to rebound like foam bedding. At some point the uphill path to the Atlantic leveled out and began to run downhill. Water from those vast freshwater reservoirs sought and found a path to the ocean.

The Great Lakes are some of North America's greatest jewels, a world unto themselves, boasting all the features of a marine environment but with freshwater ecology. Beaches, tides, Biblical storms, classic songs written about those storms: the Great

Lakes have them all along with thousands of islands, many of which are large enough to feature their own lakes. Together the five titans (from west to east: Superior, Michigan, Huron, Erie, and Ontario) represent the largest fresh bodies of water on the planet by surface area.

All the Great Lakes are naturally connected. Water flows downhill from the highest, westernmost lakes (Superior and Michigan) into Lake Huron, then south to Lake Erie. The flow wraps around the elevated headlands of southern Ontario, running east and then north over Niagara Falls to Lake Ontario where it feeds into the roaring maw of the biggest river in North America: the Saint Lawrence, essential to the histories and development of two modern nations and more than a few Indigenous ones.

The river's course tracks the boundary of the Canadian Shield, a region of ancient volcanic rock thrust above the surrounding landscape and stripped largely bare by the glaciers' retreat. What was once a landscape of tall, craggy mountains encircling Hudson Bay has been worn down by eons to rolling hills and pine forests that thrive in the depleted soil. The Shield stands apart from the Appalachian Mountains, and between them winds a deep valley through which water naturally runs to form the St. Lawrence. Its north and south banks display starkly different geology, the north rocky and scrubby while the south grows lush.

Indigenous peoples speaking Iroquoian and Algonquian languages had thoroughly settled the region when Basque whalers first chased some desperate quarry into what is now called the Gulf of St. Lawrence, the wedge-shaped inlet where the great river empties into the Atlantic Ocean. The Basques, a fiercely independent people hailing from the Bay of Biscay (straddling France and Spain), were the era's premier mariners whose exploits on the open ocean inspired western Europe's imperial powers to invest heavily in their navies. In the New World the Basques established fishing settlements along the

Gulf to provision their crews. Encountering the locals, they began trading and even established an Algonquian-Basque pidgin language to facilitate relations. The practice continued for more than a century until political changes in Europe further marginalized the Basques, and the Spanish King Philip II took control of their legendary fishing fleet. By the 1700s this fascinating hybrid language was already extinct, though, of course, nobody ever called it his mother tongue.

The Virgin, Jacques Cartier

All of which is to say that these vast territories on the far side of the sea were well known if largely uncharted by 1534 when King Francis I of France funded an expedition. To lead the voyage Francis chose a man named Jacques Cartier, who had successfully returned from voyages to Newfoundland and Brazil. Cartier was a skilled mariner from modest origins who'd married well in his home city of Saint-Malo and leveraged those connections into a stage for his grand ambitions. A pragmatist and survivor, Cartier cut a chaotic figure in his New World adventures. His brace of ships stormed the Atlantic in just twenty days, sailed down the west coast of Newfoundland and made their way to the eastern shores of modern Quebec.

Landing at Chaleur Bay on July 7, 1534, Cartier and his men encountered Indigenous Mi'kmaq people and traded peacefully with them. Sailing north he made contact with a fishing party of Haudenosaunee people from the west, upriver, who were friendly until Cartier decided on a grand gesture. He had his men construct a giant cross thirty feet high, into which they carved *Vive le Roi de France!* before sinking it fatefully into the loose, wet sand. Despite Cartier's conviction that he treated with ignorant savages, the Haudenosaunees clearly understood what it meant. Their leader, Donnaconna, approached the French, berating them for their disrespect. The ensuing standoff has been described only by French sources, who naturally portray themselves in a favorable light, but it seems to have been a sort of hostage negotiation. In the end Donnaconna agreed that Cartier could take two of his sons, Domagaya and Taignoagny, along with him provided they were returned in a year's time. While the Haudenosaunees had a strong interest in establishing trade with the Europeans, this obviously smacks of a deal made under duress.

Be that as it may, the two young Americans do not seem to have been mistreated. On the voyage back to France they told Cartier about their home settlement of Stadacona on the banks of a vast river, of the cornfields and plentiful game surrounding it, of a land to the northwest where gold and precious stones glittered in the very rocks. They called it the Kingdom of Saguenay, and Cartier found it intoxicating. Far-off lands replete with riches are common in mythology, and Saguenay was never real, but Domagaya and Taignoagny had every incentive to please their captors, and Cartier took all they said as truth.

The French court was very impressed. Eight months after landing in his homeland Cartier was off once again to the New World, helming a fresh fleet of three ships and 110 men. They left France in May 1535 and reconvened off Newfoundland at the end of July before sailing west for the mouth of the river Domagaya and Taignoagny had described. August 10 saw Cartier name a minor inlet for Saint Lawrence, whose day it was— just a bay, nothing important, but later French cartographers would name the entire gulf and river complex the St. Lawrence.

The Indigenous youths knew the river well, having spent their lives on its currents and escarpments, and it is hard to imagine what Domagaya and Taignoagny must have felt making their homecoming at the gunwales of a French warship. Cartier's fleet worked upriver until they reached Stadacona, a town of several hundred people located along the riverbank near where its mouth emptied into the gulf. The Haudenosaunee word for such a settlement was *kanata*, from which Cartier derived "Canada." Donnaconna welcomed his sons' return but had clearly learned from their prior encounter; he would not allow the youths to accompany Cartier any farther unless the French furnished their own hostages as collateral. The intrepid explorer declined (giving the game away) and continued up the river on his smallest ship, *La Petite Hermine*, until he reached the great city of Hochelaga.

That section of the St. Lawrence yawns wide to accommodate the inflow of the Ottawa river from the northwest. Planted like a stopper in this titanic flow is a large, heart-shaped island where the city of Montreal now stands. In the sixteenth century that island hosted one of the region's largest settlements: more than two thousand Haudenosaunee people who raised crops in the island's rich sedimentary soil, fished in the river and traded their surpluses with neighboring nations. Hochelaga was one of the most important Haudenosaunee sites, a vast repository of cultural knowledge, so naturally Cartier spent only a few hours ashore making perfunctory contact with the locals before returning to his ship. He attempted to take *La Petite Hermine* farther upriver but met formidable rapids on both sides of the island and could not continue. Despite his mission to seek out and secure the continent's riches, Cartier declined to send a party overland. He was a mariner, after all, and preferred not to leave his comfort zone. Cartier took his men back with the current to Stadacona.

The French fared poorly in the following months, caught completely off guard by the savage Canadian winter. Cartier knew they were roughly at the latitude of southern France and expected a mild winter but failed to account for the Mediterranean's warmth versus the frigid inland expanse that now confronted them. The St. Lawrence froze over, locking Cartier's precious ships in ice. The expedition rapidly fell ill from scurvy and might have collapsed completely were it not for the intervention of the Haudenosaunees, who gifted their stricken visitors a concoction made from a tree Cartier recorded as *anedda*, likely a spruce or cedar. Twenty-five Frenchmen died that winter. When spring finally unlocked the two-meter bulwark of river ice, the remaining eighty-five men made preparations to depart. They kidnapped Donnaconna, his sons, and a handful of other Haudenosaunees from Stadacona and made their painstaking way down the great river now treacherous with May melt. By July 1536, Cartier's ships were back in France. He presented his captives to the French court, who were duly impressed with their tales of the Kingdom of Saguenay and furnished them with material comforts that perhaps made partial amends for their circumstances. In the years following, all the captured Haudenosaunees would die in Europe. Donnaconna's good faith cost him far too much; he never saw his home again.

War in Europe drained France's coffers and delayed Cartier's return until August 1541, this time as the second-in-command of a five-ship expedition helmed by the Huguenot nobleman Jean-François de la Roque de Roberval. Cartier chafed at the demotion, though he had clearly proven himself a better mariner than a manager. Those ships carried fifteen hundred colonists along with livestock and two years of supplies, intended to establish France's first permanent settlement in Canada. Cartier sailed ahead of Roberval to Stadacona, where the Haudenosaunees made a show of numbers that the French rightly interpreted as hostility. They sailed farther upriver to establish their base, at the foot of a rocky promontory whose face glittered with crystalline stones in the sunlight. Cap Diamant, as they named it, finally yielded the treasure the French sought: diamonds hewn from the rocks and chunks of gold ore easily panned from the riverbed.

Encouraged, Cartier led another expedition upriver to Hochelaga. As before, he left the region's cultural jewel unexplored and basically undocumented to pursue the myth of Saguenay. As before, fierce rapids thwarted him just a few miles past Hochelaga, and Cartier abandoned the enterprise rather than continue overland. If a problem couldn't be solved from the helm of a boat, Cartier had nothing for it. He returned to his base for the winter to find that his men had managed to further provoke the people of Stadacona by stealing from their harvest, and now the Haudenosaunees stalked the woods in lethal bands. Winter came on as harsh as ever in that part of the world, and the colonists were as blindsided as Cartier's first expedition had been. *Anedda* kept scurvy deaths to a minimum, but the early crops the colonists planted froze in the ground, and they could no longer beg the Haudenosaunees for help. It was a miserable winter.

By June of the following year, 1542, Cartier had seen enough. He gathered up his most trusted men along with the diamonds and gold they had collected and sailed down the St. Lawrence to the sea. Just his luck: off the coast of Newfoundland he happened to meet Roberval's much-delayed fleet, finally making their way west. Roberval ordered him back to Cap Diamant, but Cartier had no intention of spending another winter in Canada. In the dead of night he stole away with his ship, hightailing it to France secure in the knowledge that it would be a year at least before the Crown uncovered his deception—by which point, Cartier assumed, he'd be amply rewarded and safe from reprisal.

Cartier's ship landed at Saint-Malo in October 1542. He presented his gold and diamonds to the court of King Francis, whose assayers swiftly determined their value to be almost nothing: the "gold" was pyrite, the "diamonds" mere chunks of quartz. Cartier's embarrassment inspired the French expression *faux comme les diamants du Canada*, "false as Canadian diamonds," a phrase so mean and funny that only the French could have coined it. Cartier found no more patronage forthcoming from the French crown, though he wasn't missing out on much: Roberval also failed to acquire anything of great value, and the Cap Diamant colony was abandoned after one more winter. The French would not return in force to Canada for another half century.

The Chad, Samuel de Champlain

The man who would actually bring French civilization to the New World came from circumstances quite similar to Cartier's. Samuel de Champlain was raised by a family of mariners in a French port city (Brouage, on the Bay of Biscay some 350 kilometers south of Saint-Malo), and, from a young age, he was trained on the sea. The young Champlain served in the army of King Henry IV before his uncle offered him the opportunity to voyage abroad.

From 1601 to 1603 Champlain served as a court geographer, meticulously interviewing sailors and trappers who returned from Newfoundland and Nova Scotia. He learned about the coasts, the inland reaches, the harsh winters, and enigmatic natives, striving at all times for both detail and accuracy. In 1603 and 1604 he voyaged to the New World himself to make detailed maps of the North American coastline. In 1608 he persuaded the King to finance a new colony in Canada. It was a small expedition—just thirty-two colonists—but Champlain had total autonomy to site, plan, and build any future he could imagine.

He chose Cap Diamant, already mapped and prepared by Cartier's expedition. A narrow stretch of the St. Lawrence River gouged out a deep bed in that spot, which kept the waters calm and navigable by ocean-going vessels. Champlain named the settlement Quebec, derived from the Algonquian word for a narrow channel or strait. He took intense interest in the Indigenous tribes and their politics, understanding he would always be vastly outnumbered in Canada, and his success would thus depend on native indulgence. Having survived his first winter, Champlain established a lucrative trading relationship with the inland nations: they brought him sumptuous animal furs in exchange for steel cutting implements and other European goods. He soon found himself recruited into a military coalition with these trading partners—the Algonquians, Wendats, Innus, and Etchemins—against the Haudenosaunees, Mohawks, and other coastal nations that began to interfere with their rivals' fur trade. These nations had already fought the Haudenosaunees for generations and sought to use the Europeans for leverage in a new offensive.

The French were few in number, but in the intimate exchanges of Indigenous warfare, their firearms could be devastating. Champlain's men won a string of victories culminating in the Battle of Sorel on June 19, 1610, where French arquebuses behind Wendat and Algonquian warriors laid a large force of Mohawks to waste and silenced hostilities with that tribe for twenty years.

The French King Henry IV was a Huguenot—baptized Catholic, the state religion of France, but raised Protestant by his mother. He remained Protestant even after ascending to the throne in 1589, though he made a show of embracing the Catholic faith to shore up his legitimacy. A pragmatic and benevolent ruler by the standards of his day, Henry IV's position bestriding the realm's principal religious divide guaranteed ample enemies on both sides. He was the target of constant assassination attempts, one of which finally succeeded in 1610. The news crossed the Atlantic in weeks, and Samuel de Champlain immediately organized a voyage back to France. With his principal benefactor dead, Champlain knew the Court would reorganize. Shortly after returning to Paris, Champlain married the daughter of the Lord Chamberlain of France.

In 1613 Champlain was appointed lieutenant general of New France, dispatched to Quebec with fresh ships, fresh colonists, and fresh dispensation to pursue France's interests. He journeyed to Hochelaga and up the Ottawa River, reestablishing fur trading relationships as he went. Champlain established a coalition of French merchants to act as a sort of investment bank, pooling their resources to finance expeditions up and down the North American coast.

Champlain understood his colony was just one stitch in the

Canadian tapestry, and thus its survival relied on good relations with the locals. These were people with whom to negotiate, not savages to conquer. His military adventures were always undertaken alongside Indigenous allies. Reunited with his old allies the Algonquians and Wendats, Champlain took ten French soldiers along with three hundred Indigenous warriors to the south shore of Lake Ontario, where they sought, found, and attacked a fortified Oneida settlement. The tall wooden palisades proved too steep an obstacle, and the coalition retreated under a storm of hissing darts. Champlain caught two arrows in his leg.

The French soldiers returned to Quebec, but Champlain, who could not walk, spent that winter recuperating with the Wendats. His information-gathering habits kept him occupied during this time, as he closely observed the tribe and conducted extensive interviews with Wendat men about their lives and beliefs. He was allowed minimal contact with the women, who held primacy in numerous areas of Wendat society such as the day-to-day management of the multifamily longhouses in which they lived. Clan organization was matrilineal. Champlain only got half of the picture, but his earnest pursuit of Indigenous knowledge stands in marked contrast to the interests of men like Jacques Cartier—for whom the shining jewel Hochelaga had been but a minor landmark.

Champlain recovered his strength in sufficient time to embark with the Wendats on their winter deer hunt, which gave him opportunities to document North American flora and fauna in great detail. Spring's melt saw Champlain ranging more widely in Wendat territory, eventually meeting up with the Franciscan missionary priest Joseph Le Caron for a diplomatic tour of the Petun, Nipissing, and Ottawa nations. Champlain returned to the Quebec colony on July 11, 1616, where he learned that machinations back home threatened to undermine his position.

This was a tumultuous time for France. Marie de Medici,

queen regent, hailed from Italy and held sway over a court that thought the teenage Louis XIII's succession overdue. They distrusted her principal advisor, who was also Italian, and pushed young Louis to take power. In response Marie imported severe rules of protocol from the court of Spain, which did little to quiet complaints of excessive foreign influence. Parisian politics were a vicious game, and absent players were dealt out, as Samuel de Champlain would discover. He returned to Paris expecting a hero's welcome, only to find that he had been stripped of his formal authority in New France.

The queen regent had no use for Champlain; his patronage, such as it was, lay with the young King Louis. Events broke Champlain's way in the summer of 1617, when Marie was pushed into exile in a palace coup led by Louis's advisors. By 1620 state affairs had settled sufficiently to let Champlain reassert his claim to authority. The King restored him as lieutenant general of New France but with a painful condition: no longer was Champlain to range the Canadian wilderness, traveling and treating with the Indigenous nations. He'd become too valuable to get shot again.

Champlain returned to Quebec and devoted himself to governance. For nearly a decade he worked at expanding and fortifying his little town on the St. Lawrence River into the biggest and most important settlement Europe maintained on the North American continent. It was so valuable, in fact, that King Charles I of England sent a fleet upriver in 1629 to take control of it. The English did not have the numbers to directly assault Champlain's fortifications, but the St. Lawrence was Quebec City's only avenue for supplies from Europe, and a blockade across the river's mouth cut them off completely. Champlain held on until his gunpowder was exhausted and finally surrendered the city.

Taken to London as a prisoner, he refused to abandon his people. As a Frenchman, if Champlain could not prevail by force of arms, he was determined to at least win the argument. To the crowns of France and England he made his case that the Treaty of Susa, signed three months before the surrender of Quebec City on the other side of the Atlantic, rendered the English conquest illegitimate. Of course nobody in North America had known about the treaty at the time, but that was neither here nor there—the treaty had to be honored. The diplomats agreed. In 1632 Quebec was returned to French control on this legal technicality with Champlain once again serving as Lieutenant General, a position he held until his death in 1635 at sixty-eight years of age. The Father of New France, the Father of Acadia, one of the greatest explorers Europe ever produced, was laid to rest in Quebec City at the heart of his life's work.

French fur trappers proliferated throughout the continent; French farmers plowed the fertile farmland surrounding the St. Lawrence. The river was their lifeline to Europe, whose people bought their grain and shipped back ancestral comforts. The banks of the Great Lakes grew thick with port towns, with European civilization, with the money and goods those connections brought. The city of Montreal, on the site of great departed Hochelaga, grew even larger than Quebec City as, for centuries, it remained the highest port on the river navigable by oceangoing ships. The Lachine Rapids, the perpetual eight-foot standing wave that turned back Jacques Cartier time and time again, would remain essentially undefeated into the twentieth century when the St. Lawrence Seaway project opened all the Great Lakes to oceangoing vessels. The St. Lawrence River was life itself to the First Nations and life itself to the Europeans who paved over their bones.

Mississippi Basin

The Spanish and the English distinguished themselves as aggressive colonizers from the moment they set foot on the North American continent, swallowing up as much territory as their troops could defend against the Indigenous population. They sought gold and farmland, and the locals could only get in the way. The French adopted a policy of coexistence, not due to moral or humanitarian concerns but rather based on the conviction that such an approach was better for business. The economy of New France was based on purchasing top-quality animal pelts from Indigenous peoples; killing or alienating them would ruin the whole enterprise. These competing philosophies of empire would be put to the test in the slow, murky waters of the Mississippi River. Nowhere else in North America does so much land drain down a single course. The Mississippi has expanded and contracted repeatedly, but for the last seventy million years it has channeled the waters of the American Midwest to the Gulf of Mexico. From the eastern slope of the Rockies to the western slope of the Alleghenies, nearly every river and stream pays tribute to the "Father of All Rivers."

The first European to lay eyes on the Mississippi was the Spanish explorer Hernando de Soto. At the head of a large and well-provisioned expedition, de Soto marched north and inland from Florida in 1539 before venturing west across the Appalachians. De Soto carved a bloody path through the landscape. His men stole from Indigenous Americans, as well as killing and enslaving them. He scouted a large swath of the American South before finally sighting the Mississippi River on May 8, 1541. At that time the river was heavily populated by dozens of Indigenous nations, some of whom lived off fish and game while others erected permanent structures and practiced agriculture. It would probably be incorrect to say the Mississippi teemed with traffic, as the mammoth river would have dwarfed the shallow-keeled Indigenous boats, but this was not *wilderness* in the classic American conception. It was the heart of a thriving civilization.

Having insisted to the locals that he was a god, de Soto must have been gravely embarrassed when a mere fever struck him dead on May 21. Hungry, with half their force dead, their gunpowder spent and crossbows corroded, de Soto's lieutenants began an agonizing retreat south toward the Texas coast. Europeans would not traverse the river again for more than a hundred years.

The French took their turn starting in 1673, led by a Jesuit priest and a seminary student turned fur trader. Father Jacques Marquette had heard tales from the local Indigenous people of a vast river, far to the southwest, that served as a highway for people and goods. He partnered with trader Louis Jolliet, and together these men and five others carried two canoes overland from the Fox River, near Lake Michigan, to the Wisconsin River just two rough miles away. Despite their close proximity the rivers flowed in opposite directions: the Fox ran north, toward the lake, while the Wisconsin ran south, away from it. This quirk of geography is the result of the so-called Niagara Escarpment—a plateau of rock elevated above the surrounding land, creating a curved ridge for a thousand miles around the edge of the Great Lakes. Water collects in the basin it forms and avoids the ridge, daring to cross it only at the massive precipice of Niagara Falls (whose constant impulse gives the flow enough speed to vault the ridge). On the north side of the ridge lies the St. Lawrence Basin, draining to the North Atlantic; to its south the Mississippi basin drains to the Gulf of Mexico. Marquette and Jolliet had found a sort of continental divide.

The two canoed down the Wisconsin River to the Mississippi, documenting their journey all the way. The river they found was exactly the highway of man and nature they had heard described. Several hundred miles south, the Ohio River joined the Mississippi and the flow became truly staggering, so wide that in some sections the water seemed to stretch from one horizon to the other. Aiding the impression was the flatness of the territory itself—from Minneapolis, near the Mississippi's headwaters, to New Orleans, at its mouth, the river runs two thousand miles while only dropping 830 feet. The water runs slow and shallow at first, picking up depth and speed as tributaries add to its volume. The Indigenous peoples Marquette and Jolliet encountered were friendly and eager to trade, easing their journey south. In just two months they'd reached the mouth of the Arkansas River, where they turned around so as not to encroach on Spanish territory. A quick survey was all it took to convince them of the Mississippi's tremendous commercial potential, from trading furs with locals to farming in the basin's rich sedimentary soil. Marquette and Jolliet returned to Wisconsin in September 1673, and their reports would inspire the next generation of French explorers.

By the end of the seventeenth century, French traders had begun to make contact with the Indigenous peoples of what the colonizers called Louisiana. They located the mouth of the Mississippi and founded the settlement of New Orleans there in 1718, setting it on a hill above a bend in the river where it might survive the Biblical storms and floods that rode in at the end of each summer. French trading posts sprung up along the banks of the Mississippi, the Missouri, the Arkansas, the Ohio, and the Illinois Rivers before Napoleon Bonaparte ultimately sold the territory to the United States in 1803.

Navigation was usually a one-way affair for Europeans on the Mississippi. Floating cargo down a river was one of the cheapest possible ways to move goods, but going upriver was a grim prospect. Typically one paid a crew of strong-backed men and livestock to haul one's boat along the riverbank with heavy ropes the French called *cordelles*, but this was so expensive that few merchants bothered. Goods generally traveled north to south, borne along the vast river aboard cheap little craft with flat bottoms and shallow drafts to avoid submerged debris. Seasonal storms introduced dangerous new obstructions to the riverbed every year. Hauling anything heavy was dangerous to both men and capital. When the flatboats reached their destination, they were dismantled with prybars and sold for scrap. Nobody who dropped a flatboat into the current expected to get it back. They could only hope its cargo arrived intact.

The American Age

When the United States of America took possession of the Lou-isiana Territory, it claimed the Mississippi River in its entirety. The young nation thrummed with energy; the Industrial Rev-olution and waves of eager European immigration presented stiff challenges to an adolescent American government. A rap-idly expanding population competing for limited resources threatened unrest, but Thomas Jefferson had an answer in the prospect of westward expansion.

Jefferson imagined the ideal American citizen as an indepen-dent farmer: the freeholding master of a plot of land, empow-ered to work, develop, and enrich that plot to the limit of his resources. The nation he envisioned was agrarian, a patch-work of independent farms overseen by a limited government devoted primarily to adjudicating property disputes. In the Louisiana Territory Jefferson saw an opportunity to make mil-lions of landless, disenfranchised Americans into a new gen-eration of yeoman farmers. He would buy off the citizens of this fractious new country with vast tracts of the world's most fertile farmland, disbursed by the federal government at prices ranging from outrageously cheap to literally free. This policy represents the earliest manifestation of what later genera-tions would call, with gradually escalating levels of irony, the American Dream.

In 1811 the steamboat *New Orleans* became the first steam-powered vessel to make the journey from Pittsburgh down the Ohio River to the Mississippi, and thereafter to its name-sake city on the Gulf Coast. Sporting the twin smokestacks, steam-driven paddlewheel and flat bottom we've come to asso-ciate with an entire era of American history, the *New Orleans* could chug upriver almost as easily as it floated down. North American commerce re-ordered its constellations overnight. In 1814 the Port of New Orleans listed twenty-one steamboats on its call manifest; five years later 191 were listed, and by 1828 that number was over 1,200. Soon it became cheaper to ship goods from Cincinnati to New York City via the Mississippi and the Atlantic than it was to truck them one tenth of the distance overland.

In the early 1800s a ship could sail from the western tips of Lakes Superior or Michigan all the way through Lakes Huron and Erie unimpeded. But reaching Lake Ontario was impossible due to the fact that the waterway drops right over the Niagara Escarpment to form Niagara Falls. To skip the waterfall ships would dock in Buffalo, a port at the Niagara River's mouth. Cargo destined for Boston, New York, or Europe could be hauled by mules downhill to the shores of Lake Ontario, reloaded onto riverboats for a jaunt down the St. Lawrence to Montreal, offloaded a second time on the western side of the Lachine Rapids and reloaded a second time onto oceangoing ships in the great port on the east side of the island. All these steps added delay and expense to the enterprise—the journey from Buffalo to New York City measured more than a thousand miles at sea, crossed the U.S.-Canadian border twice, involved a minimum of three ships and two wagons, and was unavailable during the winter due to river ice. The United States badly needed an internal link between the Great Lakes and the eastern seaboard cities. In response to this need the State of New York drew up plans for the most ambitious public works project in the young nation's history: a canal linking the port of Buffalo with the town of Albany near the top of the Hudson River 350 miles away, down which cargo could be easily barged to New York City.

Thomas Jefferson saw no role for the federal government in internal improvements. Jefferson dismissed New York's plans in a tone that would grow familiar to advocates of public spending throughout American history: "It is a splendid project and may be executed a century hence . . . here [in DC] is a canal of a few miles projected by General Washington which has languished for many years because a small sum of

$200,000 cannot be obtained . . . think of making a canal 350 miles long through a wilderness! It is a little short of madness to think about it."[2]

Jefferson successfully discouraged the New York officials, but by 1816 economic imperatives pushed the state to fund the Erie Canal by itself. They planned the route to include eighty-three "locks": chambers sealed by sluice gates that act as elevators for boats, allowing canals to cut through areas over which gravity would not naturally allow water to flow. Teams of horses pulling towropes would provide the locomotion to move ships up and down the canal. A fine plan but, even with funding, the state still had to find people to build it. The United States had no tradition of civil engineering, no centralized apparatus to train or accredit the men (exclusively men, at this time) who might actually do this important work. European monarchs had been sponsoring academies and ministries for generations, the better to break down the work of governing a nation of millions into consistent results-driven processes and to amass a collective store of expertise for future works. In this sense, nations like France rationalized their governments before they democratized. The United States developed in precisely opposing fashion, like an embryonic octopus growing its eyes from the outside in—we democratized before attempting to rationalize. Thus the men responsible for building the Erie Canal would not be trained experts but rather enthusiastic amateurs.

New York's government wanted to hire an Englishman named William Weston, who was considered one of the best canal engineers available by virtue of being one of the only European-trained engineers on the continent. As historian Elting Morison wrote, "Knowing not much, he knew a great deal more than anyone else and was in frequent demand."[3] Weston's projects had a tendency to fall apart a few years after completion. He inadvertently did everyone a favor when he bowed out due to political considerations (the War of 1812 was ongoing). The state turned to Weston's sometime business partner, a judge and professional surveyor named Benjamin Wright. Another principal engineer, James Geddes, was also a judge and surveyor. Surveyors served a vital role in the rapidly developing nation, where they resolved arcane disputes over property lines. The state hoped trained surveyors could do the detail-oriented work of mapping elevation changes over hundreds of miles. Wright and Geddes learned as they worked, hiring scores of junior engineers who knew barely less than their bosses and enjoyed wide autonomy on the job once real work began in 1817.

A young man with the fabulous name of Canvass White became the most successful of their acolytes. He studied mathematics and surveying but, like the others, had no formal engineering experience. Wright tasked him with designing waterproof locks for the canal; American locks were built with wood, which inevitably rotted and needed replacing at great expense. European engineers had developed "hydraulic cement" that reacted with water to become impermeable, but it was too expensive to import at the scale required. White paid his own passage to Europe, where he learned from the world's best canal engineers how the cement was manufactured. When diggers uncovered beds of pristine limestone along the canal's route (the Niagara Escarpment is made mostly of limestone) White saw an opportunity. By the spring of 1819, when the snow melted and men could work again, White had perfected his hydraulic cement formula using American limestone, and, following the canal's completion, White would find himself the nation's premier cement engineer. White's colleagues elsewhere on the project invented new methods for downing trees, digging up stumps, and shredding subterranean roots. Mathematically inclined young men flocked from across the country to join the effort. The Erie Canal was in many ways the United States' first academy of engineering. Nearly every renowned American civil engineer of the nineteenth century either worked on the Erie Canal or received his training from someone who did.

When the canal needed to cross the Mohawk River valley, characterized by steep hills that ran all the way to the riverbank, Canvass White decided to run the canal down the riverbed itself for a stretch before hopping it over to the north bank on a 748-foot aqueduct. Twelve miles later it went back to the south bank on a 1,188-foot aqueduct in order to take advantage of a smoother downslope. To the west, near Rochester, the Irondequoit Creek presented its own obstruction. The creek was small but old enough to have cut itself a deep valley that the engineers did not want to build locks in and out of, so they decided to connect three natural ridges by means of two artificial ridges: 1,500 feet long and built of compacted earth. When local soil proved unsuitable for this purpose, they hauled in thousands of tons of dirt on mule carts. To conquer the Niagara Escarpment they hacked five sets of double locks into a limestone cliff, removing 1.5 million cubic yards of rock in the process.

Work on the Erie Canal finished in 1825, though sections of it opened earlier. Suddenly the journey from Buffalo to Albany, which had taken two rattling weeks by wagon or stagecoach, became an effortless five-day glide by boat. Freight rates plummeted 90 percent. The canal's impact on commerce and culture across the state of New York cannot be overstated: it came to be called "Mother of Cities" due to all the settlements that sprung up along its path. Dragged by animals walking beside the canal, frequently stopping while locks filled and emptied, canal boats moved slowly enough for passengers and crew to patronize canal-adjacent merchants without disembarking. Money poured into the state's underdeveloped interior. The state of New York spent almost $8 million building the Erie Canal and recouped that cost from canal tolls within a decade. Meanwhile, New York became the nation's most important commercial channel—by far the cheapest and safest way to move people and cargo between the Great Lakes and the East Coast. Buffalo became one of America's most prosperous cities, and New York City surged past Philadelphia as the United States' biggest and most important seaports. Immigrants from Europe took advantage of the cheap passage and abundant farmland in the Midwest to ride the Erie Canal toward their futures, depositing their cultural artifacts like seeds along the way.

Canal projects exploded across the country, most of them helmed by men who'd made their names on the Erie project. In just eight years of work Benjamin Wright, James Geddes, Canvass White, and their colleagues had not merely built a vitally important system of infrastructure, not only established an American tradition of engineering where none existed before, but they made credible the notion of a government spending heavily on internal infrastructure to stimulate economic development. Not bad for a crew of amateurs.

The Destructive Impulse

Rivers meant life, agriculture, commerce, and riches to Indigenous and colonial Americans, but, prior to the twentieth century, they were also exceedingly dangerous. The steamboat trade, massive as it became, was beset from first to last by horrific disasters claiming hundreds of lives. Often the hazards lay in the rivers themselves: logs, stumps, boulders, prior shipwrecks, and other obstructions hid below the water's surface and could not be mapped as currents dragged them about the riverbed. Steamboat wrights built their ships with flat bottoms for shallow draft, riding just a few feet below the water line, but every voyage carried a real element of risk. A single obstruction could impale the ship's soft belly and send the whole mass to the river bottom, where it might be buried in silt and never recovered. Recognizing the threat river snags posed to interstate commerce, the U.S. Army Corps of Engineers began dredging sections of the Mississippi River as early as 1810. By 1832 they were dredging the entire main channel of the Mississippi as well as the lower Missouri River. States did the same work for important navigable rivers within their own borders. Every storm, flood, or shift in the river's path created new work for the dredgers, who performed tough labor in dangerous conditions to keep cargo and passengers safe on the waterways. In 1882 a single Corps of Engineers snag boat removed over twelve hundred obstructions from the Red River.

Even as the nation's waterways became less hazardous, the profit motive spurred new dangers. Enterprising shipwrights assembled thousands of steamboats in just a few decades, most learning as they went, and they often did their work on the cheap. The typical "golden age" steamboat consisted of a lightweight wooden frame slathered with flammable paint and varnish, yoked to a set of enormous steam boilers that drove the paddlewheels and sent plumes of coal smoke out the iconic double stacks. The boilers burned coal to vaporize river water, but steam is inherently dangerous under pressure—a contained explosion of potentially lethal force. "Fire tube" boilers, which recycle superheated exhaust to boil water with less fuel than a conventional boiler, appeared in 1848, offering dramatically increased efficiency at the cost of greater risk. If conditions in the machine were not precisely maintained, pockets of excessive heat could form that critically weakened the boiler's casing and allowed steam to escape. Many steamboats burst into flames following boiler mishaps, and a few simply exploded.

The steamboat *Sultana*, launched in January 1863, at 260 feet in length with a draft of seven feet, undertook her final voyage in April 1865. The Civil War was coming to an end after nearly four years, though the violence would persist for years— President Lincoln had just been assassinated and unrepentant Southern elites plotted to force Black Americans back into de facto slavery. Nevertheless with the Confederacy's surrender, both sides began repatriating their prisoners of war. The *Sultana* steamed south to Vicksburg, where her captain accepted a bribe to carry an entire camp's worth of Union parolees to the

North. Though *Sultana* was meant to carry no more than 376 passengers, she found herself carrying nearly two thousand.

Slow rivers tend to wind, to form dramatic "oxbows" as sediment accumulates into mounds and redirects the flow, and the preindustrial Mississippi was littered with bends the federal government has since straightened and dredged. As the overloaded *Sultana* wheezed her way upriver, through a fast and energetic spring melt that left treetops poking above the water's surface, each port and starboard turn caused the ship to list badly in that direction. The rolling motion caused the water in her four boilers to slosh from side to side, upsetting the balance in water levels necessary to keep them safely operating. On the night of April 27, 1865, three of her four boilers exploded almost simultaneously.

The blast traveled upward, demolishing a large chunk of the ship's structure, obliterating the pilothouse and setting the *Sultana* adrift in the river. Hundreds died in the first instant; hundreds more in the awful minutes that followed, as the twin smokestacks toppled over to crush soldiers who crowded on every deck and surface. Flames raced through what remained of the structure. Anyone who could make it to the water leapt overboard, but these men had endured months, if not years, of malnutrition and deprivation at the hands of their Confederate captors. Most were too feeble to swim, if they knew how to swim at all. Weaker men clung to stronger men, and the river claimed them all.

Farmers along the Mississippi's banks saw the *Sultana* burning in the dark, a scene straight from perdition. Striking out in boats and log rafts, they scooped up survivors and deposited them in the tops of trees still exposed above the flood. More than a thousand men died that night—the worst maritime disaster in American history—but the nation was too consumed with tragedy to notice, and the *Sultana's* fate received little press. These men had fought to save their country and end the monstrous evil of American slavery, had suffered in atrocious conditions endemic to nineteenth-century Prisoner of War (POW) camps. They should have been sent home as heroes; instead, an avaricious merchant literally sold them up the river, packed them in like rats, and left them to die. Captain James Mason, the disaster's principal architect, died in the initial explosion. This made it easy for investigators to let the Union quartermaster who did the bribing, Army captain Reuben Hatch, evade responsibility. Hatch's older brother, Ozias Hatch, was one of President Lincoln's close advisors, and nobody was eager to punish him.

The Mississippi, harboring so many cities along its banks and draining so much of the country, regularly killed hundreds of people at a time. Steamboat disasters accounted for just a small fraction of the casualties, as Biblical floods were regular events on the river. Every spring the snowy Upper Midwest sends a minor ocean of fresh water down the Mississippi. The land around the river is flat and the typically slow-moving water has never run between high riverbanks, so they're easily overtopped. The Mississippi, in its preindustrial state, was a plastic organism: swelling and shrinking month to month, constantly shifting over the landscape, carving new channels for itself on its relentless southward path. Where the water decided to move, human settlements were simply swept away. Indigenous Americans avoided building permanent structures on the Mississippi's flood plains for precisely this reason, choosing instead to migrate between low-lying summer grounds and elevated winter grounds. Staying put by the riverbank in flood season would have struck them as suicidal. European colonists, on the other hand, organized their societies around private property rights. Every acre of land was owned by someone, and each farmer was only permitted to work the plot of land allotted to him, even if that land was literally underwater. For this principle the Europeans and their Americanized descendants were willing to spend a bottomless store of treasure and sacrifice a limitless number of human lives. The river flooded every spring, some years worse than others, and the people close to the water just had to try and survive. Their homes, their businesses, and their cropland were often all they possessed, and departing would render them destitute.

Because reordering the values of their society seemed out of the question, nineteenth century Americans applied their labors to what seemed a likelier prospect: controlling the river. In 1879, with the Confederacy vanquished and the South badly in need of development, the U.S. Congress established the Mississippi River Commission (MRC) to make improvements to the river in order to foster navigation, promote commerce, and limit destruction from flooding. Funded and empowered to create comprehensive basinwide plans, the MRC set about building wing dams (also called wing dikes or weirs), partial obstructions sitting under the water's surface that direct flow toward the center of the river. This faster flow carves deeper channels for safe navigation and the slower flow near the banks encourages sediment accumulation, effectively shoring them up. The MRC also began digging up large volumes of dirt and piling it into levees—earthen mounds lining the banks that kept rising water in the riverbed. High enough water would overtop the levees, fast enough water would strip away the mass they needed to hold the river back, and a poorly constructed levee might get lifted up by water and washed away. So many risks, but the nation's interior depended on reliable access to river commerce. They would just have to eat the losses.

The 1880s delivered to the American Midwest some of the worst winters in recorded history, with historic volumes of snow that, come spring, rushed downriver as lethal floods. Flood damage across the country inspired citizens to protect themselves by building their own levees and even (on smaller rivers not used for navigation) full-span dams. Dams limit flood damage by pooling flood water in a reservoir behind the dam, which may be released during periods of low precipitation through tubes called penstocks to maintain open capacity in the reservoir. If a flood fills the reservoir faster than the penstocks can drain, excess water bypasses the dam by means of an emergency spillway. If these safety measures aren't adequately observed and maintained, the dam stops functioning as flood protection and becomes an active hazard to anyone and everything for many miles downstream.

Johnstown

The state of Pennsylvania first built the South Fork Dam in the 1840s in the thickly wooded hills east of Pittsburgh. A total of 72 feet high and 931 feet long, built from mounded earth and clay, it stopped up the Conemaugh River in order to establish a reservoir for a system of Erie-inspired canals running throughout the state. Once proliferating railroad lines rendered the Conemaugh canal section obsolete, the state government sold the dam and reservoir to a railroad company that sold it to the South Fork Fishing and Hunting Club. Pittsburgh's wealthiest industrialists (Andrew Carnegie and Andrew Mellon were both members) made the club their exclusive country retreat, where they could socialize far from anyone who might be jealous of their richly deserved rewards. When the dam needed repairs, the club had the work done cheaply—shoring up regular leaks with mud and straw. The iron penstock pipes had been removed and sold for scrap by the railroad company. The club enjoyed easy fishing on the reservoir, so much so that they installed a fish screen in the spillway that clogged it with river debris. Despite complaints from the downriver city of Johnstown about the dam's condition, the club left itself without any mechanism to lower the reservoir in case of flood.[4]

Events accelerated in the last days of May 1889, when a fast spring melt coincided with a mammoth storm. Six to ten inches of rain fell across the region in less than twenty-four hours. Caretakers at the South Fork Club worked to shore up the failing dam, but it collapsed on the afternoon of May 31, sending over three billion gallons of water downstream in sixty-five minutes. For one afternoon, the Conemaugh—a narrow, stony channel that needed a reservoir to maintain navigable water levels—carried the volume of the Mississippi River.[5]

The town of South Fork, immediately downriver, was close enough for its citizens to see the water pouring down the dam's face. They scrambled to high ground, abandoning their homes, knowing that an overtopped dam was a dam fatally compromised. Sure enough, water burst through the dam's face and raced downhill carrying a good portion of the earthwork along with it. South Fork was consumed, its homes and businesses smothered by the unstoppable flow. All but four residents survived, bearing witness to one of the most terrifying phenomena spawned by a wrecked dam: the materials the water picked up moved along with the flow accreted at the flood's head into a single solid mass. Eyewitnesses described the flood not as a body of water but an impossible mountain in motion, "a huge hill rolling over and over."[6] This ball of death, this Katamari of unholy destruction, continued down the Conemaugh's course until it reached the city of Johnstown, built thickly on the riverbanks.

The wall of debris and water that struck Johnstown was between forty and sixty feet high and moving at more than forty miles per hour. Nobody had any warning. The debris monster crushed and grabbed everything in its path. Just upriver from Johnstown stood the Cambria Iron Works whose boilers exploded and whose warehouses contributed mile after mile of freshly manufactured barbed wire to the flood monster. To those on the ground it would have seemed like the end of the world, sound and fury enough to shatter one's mind just before the weight of a dead steer wrapped in barbed wire crushed one's body to paste. The railyard in East Conemaugh became another grim scene, the flood tossing cars about like toys. It carried a fifty-ton locomotive for a quarter mile. Freight Conductor Fred Brantlinger jumped from his train and tried to climb a nearby slope. Brantlinger's account to authorities hints at his horror—not so much at the devastation as its otherworldly nature:

> I went for the hill, and when I would get up a piece I would slip back; the second time the water got to me . . . it raised me right up and I got hold of a limb of a tree, and as the water raised, I went along from one branch to another. It was horrible, I tell you. There was a draft of very strong air ahead of the water, that I believe was worse than the water, for when that air struck me, it seemed to lift me right off of the ground. A man could see it; it just looked like a blue heat, you know. I don't know whether it was imagination or not but I thought I could see things falling before the water got to them. It made a terrible noise; you would think the whole earth was being torn up.[7]

The "blue heat" he described was a shockwave: a bolus of air compressed by the flood's mass into such high pressure that it felled buildings and bent sunlight into new colors. The flood smashed buildings, scoured away soil and vegetation, left some patches stripped to bedrock. Twenty-two hundred people died that day. The Johnstown flood represented the worst civilian mass casualty event in the history of the United States, not to be surpassed until the hurricane-driven flood of Galveston, Texas, in the year 1900. The survivors and the families of the dead pursued the principals of the South Fork Hunting and Fishing Club with lawsuits, but courts consistently ruled in favor of the wealthy landowners. Under the doctrine of fault-based liability accepted at the time, the owners could not be blamed. It was not they who built the dam nor removed the penstocks nor caused a historic glut of rain. This obvious injustice heaped atop other high profile dam-and-reservoir disasters in Europe to alter the legal system's approach to liability. The developments that followed led to the doctrine of strict liability, holding that property owners are liable for their property regardless of fault.[8] Johnstown represented a fundamental shift for the law and the nation's perception of large dams. For the first time many people began to appreciate the staggering power behind these structures and the risk they represented to downstream communities. Under strict liability few private companies were willing to shoulder the risks associated with large water projects. Dams and reservoirs increasingly became the exclusive domain of governments.

2

BRICK BY BRICK

The American River is born each day in the Sierra Nevada. As the sun rises over California's eastern frontier, water seeps from stone and crawls down the ruddy slopes, finding its way into three distinct digits that converge with gravity and empty into a deep glacier-cut gorge. Rushing west under stony bluffs lined with towering oaks, the river ends its 120-mile sprint washing into the Sacramento River just north of the city bearing the latter's name.

Indigenous Americans settled the region thousands of years ago, but it remained untouched by colonizers until the start of the nineteenth century. Trappers slogged through lush marshes, paddling between house-sized stands of tule reeds in search of the delta's copious beavers. In 1806, men from the Mexican territory of Alta California ventured inland from their mission settlements, where the since-sainted Franciscan priest Junipero Serra kept hundreds of Indigenous people in a state somewhere between paternal care and feudal serfdom. Several natives had renounced the Padre's generosity and flown the coop, so a Spaniard named Gabriel Moraga tracked them inland with a party of soldiers. Moraga would eventually turn back from the endless marshes and understandably hostile locals without ever sighting his quarry. In a historic display of sour grapes, he named the place Río de las Llagas, the "River of Sorrows."

Famed fur trapper and explorer Jedediah Smith worked the river in the mid-1820s. His crew's notable, prolonged, and likely obnoxious presence inspired the locals—who by this point spoke Spanish as an intertribal tongue—to call it Río de los Americanos. The name never really changed thereafter, though the land Moraga and Smith navigated would be unrecognizable today. The great Sacramento Delta is gone, tamed and reclaimed and converted to farmland, pocked by barren flood plains that stretch for square miles under elevated interstate highways. The change would seem inconceivable to them, like demolishing the mountains.

Water flows from the mountains of southern, northern, and eastern California downhill toward the low-lying center of the state. In winter months the delta was an inland sea, more navigable by boat than carriage, populated by herds of elk and industrious beavers that hid from trappers in impenetrable thickets of tule reeds. Every spring and fall the sky darkened under untold millions of beating wings, as the delta has for thousands of years been a major staging ground for most of the migratory bird species west of the Rockies. It was by all accounts the Everglades of the West, and now it is more or less completely gone.

The water has not vanished. It is being piped elsewhere, into the rich peaty soil hosting fruit and nut orchards up and down the Central Valley. Three-quarters of California's fluid bounty goes to agriculture, and the state government sacrificed the Delta to make it happen. The region south of Sacramento where the Delta used to be is now a wholly artificial zone of crops, canals, and levees—among the most thoroughly engineered landscapes in the world. Whether or not anyone thinks this trade worthwhile is moot, for the choice was permanent and what they decided then they decided for all of us forever. That's the reality of the infrastructure-built West.

A Prison for a Dam

California was annexed by the United States in 1848 and received statehood in 1850, fully opening her prodigious resources to

American exploitation. Gold prospectors ravaged the land for ore, particularly along riverbeds where nuggets might be found among finely polished gravel and twinkling garnets of salmon roe. Turning the river's power against itself, they used hydraulic blasting to scour the banks of foliage, earth, and stone. The boulders and caustic chemical waste from this process they let wash downstream, permanently altering the riverbed and raising the water table, destroying generations of embryonic salmon and poisoning or drowning untold acres of cropland. Even shipping via the inland route from Sacramento to San Francisco suffered as barges ran aground on the rubble of the Sierra Nevada. The state's 1884 ban on hydraulic mining is often called its first environmental law, citing the "public and private nuisance" left in the practice's wake.[1]

Riverbed prospecting dried up after a few decades as the accessible ore ran out, but the banks of the American River nonetheless buzzed with new development. Sacramento was the western terminus of the First Transcontinental Railroad, opened in 1869. Opportunities abounded, and some twenty miles upriver in the foothill town of Folsom a wealthy family hatched a modest scheme with massive implications.

Horatio Gates Livermore arrived in California in 1850 seeking gold and secured enough of it to get himself elected to the state senate four years later. His sons, Horatio Putnam and Charles, joined him out west to help secure a large plot of riverside land in Folsom. They planned to build a diversion dam, taking a portion of the river's water to power a sawmill they also meant to construct. When it became clear the project was beyond their immediate means, the Livermores sought new funding. The family patriarch's connections in government positioned him to pounce on opportunities, and in 1868 when California needed a site for a new state prison, the Livermores were ready. In exchange for 250 acres on the American River's east bank, the entrepreneurial family bought themselves $15,000 worth of convict labor. The men who'd do that work weren't yet in the prison that would do the leasing nor had many of them yet committed the crimes that would condemn them, but somewhere in the ledger of stars their fate was inked.

Folsom State Prison opened in 1880, seventy-five years before Johnny Cash dropped his famous record. In 1881, Horatio P. Livermore (now running the family business) reorganized the holding company for the dam project into the Folsom Water Power Company. This new entity claimed its right not only to the convict labor promised twelve years earlier, but double that amount. California fought their demands but lost in court, ordered to provide sixty thousand man-days of labor annually for five years in exchange for use of the power company's railroad and a paltry volume of water from the diversion canal once completed.

It was all the free labor the younger Horatio would need to meet his upgraded ambitions. In the years since first planning the dam, he'd grown enamored with hydroelectric power, which several new dams in the distant Northeast were starting to produce. Horatio dreamed at greater scale, imagining his project powering not just a few mills in Folsom but the entire city of Sacramento. He wanted to build something never before seen in the world, let alone the West, and convict labor liberated the capital to do it right.

The diversion canal, once finished, ran water for more than two miles along the lip of the gorge while the river fell away beneath, ending in a storage basin eighty-five feet above the riverbed. On the steep slope between the two bodies of water, directly in the path of all its potential energy, Horatio built his powerhouse. For the stonework he hired masons from Italy, judging them the world's finest on the basis of Roman engineering from two thousand years prior. However questionable his reasoning, the men he hired raised a handsome three-story structure in brick and sandstone, the interior of which with its high vaulted ceilings manages to recall a Mediterranean grotto. The building hosted four penstocks: iron tubes eight feet in diameter through which thundering volumes of water fell from the forebay to the river. This relentless force moved man-sized turbine blades inside the penstocks, in turn driving four generators.

The generators were technological marvels—state-of-the-art machines built to specification by General Electric in Schenectady, New York. They still stand in situ, hulking black golems each massing nearly thirty tons, described by a contemporary journal of electric science as "without doubt the largest three-phase dynamos yet constructed." Too large for rail transport, they were instead shipped nineteen thousand miles around Cape Horn and installed at the Folsom Powerhouse in 1895. When the penstocks were opened and water roared through the system, an eight-foot ring of copper "branch circuits" whirled like a carnival ride around six cinderblock-sized magnets of coiled iron wire. Each of the four generators could produce 750 kilowatts of electricity, and of very significant importance the output was *alternating current*.

There are two types of electrical current: direct and alternating. Direct current flows through a circuit in one direction only (say, from a flashlight's battery to its bulb), while alternating current flows back and forth in a pulsing wave pattern. Direct current (DC) was the first type of electricity to find commercial employment in the early nineteenth century, but our modern world largely runs on alternating current (AC) because its voltage can be modified with the use of "transformer" equipment. Electricity is generated at one voltage, hopped up to very high voltage with a transformer for efficient transmission over long distance and then dropped back to lower voltage for commercial use by a local transformer. Our whole distribution system relies on the efficiencies AC provides. DC current is still used in common devices like cell phones and computers, that require adapters to convert electricity from the AC-driven grid into DC for use. In 1895 most generators produced DC, so they had to be built very close to where the power was needed—running a factory, say, instead of a city's power grid.

The Folsom Powerhouse was a daring attempt to change how electricity was produced and delivered in the American West. The dynamos Horatio Livermore paid to have shipped halfway around the world spun at three hundred revolutions per minute, linked by a tangle of rubber belts squealing their way around the spoked wheels of a bewildering machine called a Lombard Water Wheel Governor. The governor kept the four generators running at the same speed by using centrifugal force to limit the flow of water to any turbine that started running too quickly, and the result was current pulsing out of the powerhouse at the unerring rate of sixty oscillations per second—the same frequency pulled from power outlets across the modern United States. The Folsom Powerhouse didn't set this 60-hertz standard alone, but, as one of the West's first producers of AC power, it played an influential role.

Twelve air-cooled transformers representing the height of contemporary technology boosted the generators' output from 800 to 11,000 volts before launching the supercharged current downriver. The cables carrying it were compound in nature, bundling together some ninety copper wires into a mass as wide around as a dinner plate. They ran uninsulated atop cedar poles for twenty-two miles, bleeding energy as heat the whole way to the Livermore's Sacramento substation. The same company that owned the substation also ran a network of electric streetcars, letting the gold prospector's sons profit at every step of the process. In 1895 the capital of California was no longer a muddy, windswept prospector town but a hub of modern industry.

The streets were lit by arc lamps—the dream of every steampunker—casting otherworldly blue sparks back and forth over a half inch of empty air, their ozonic crackling drowned out by the sound of rolling rail stock. To any reasonable observer Sacramento would have seemed a perfect portrait of the rapidly developing West, at the forefront of American ingenuity. On the night of September 9, 1895, the city formally celebrated the Livermore's historic achievement with a Grand Electric Carnival. Thousands of Edison bulbs lined the wires strung like Christmas garlands from building to building and over their facades, in crossings high over smooth asphalt streets. In a photo of that night the steel of the street-car rails captures the light and winks it back at the camera, drawing the eye down the street where in the distance the lights and buildings converge to a shining pyramid: a curtain of lights tapered at the top like a gazebo, brilliant even with the separation of 120 years. Dangling from a cable in the foreground is a blazing assembly shaped like a grizzly: the Golden Bear of California, the majestic animal that in 1895 was already being hunted to extinction.

The Folsom Powerhouse forever altered the Sacramento region and everything downstream. Demand for electricity far outpaced supply in the early twentieth century, so private investors began building hydroelectric dams on waterways up and down the Sierra Nevada. The mountain range's steep drops, copious winter snowpack, and glacially acute geometry created a massive supply of power, and alternating current generators as pioneered at the Folsom Powerhouse meant they could transmit that current all the way to the San Francisco Bay Area where ports and factories multiplied. The Folsom Powerhouse's transformers underwent substantial improvements around 1918 to keep pace with new technology, but the original turbines and generators kept operating for decades. Though theoretically outdated and difficult to maintain, these black metal beasts were so superbly engineered that the Pacific Gas & Electric Company (PG&E) thought it would be wasteful to shut them down.

In the end it was not General Electric's machinery but the dam itself that met final obsolescence. The 1930s and 1940s were a fairly wet period in California, historically speaking, and the mercurial American River had a long record of destructive floods. As part of the Central Valley Project, which aimed to build a single vast system of water control from the state's soaked northern half to its parched southern reaches, a new dam was planned on the American River at Folsom. Between 1951 and 1955 the new Folsom Dam went up adjacent to the old diversion canal, transforming this modestly modified stretch of river gorge into a magisterial spectacle: a wall of earth and stone stopping the river. Horatio P. Livermore dug more than two miles of canal along the bank to build up eighty-five feet of potential energy; the Folsom Dam's penstocks have three hundred feet of hydraulic head driving their turbines to a production capacity of nearly two hundred megawatts (the powerhouse could manage about three MW). In 1952 the obsolete Folsom Powerhouse ground to a halt after more than half century of operation. PG&E sold the site to the state for a public park, where the powerhouse has been carefully preserved. The mighty generators, cast and riveted and assembled by hand over 120 years ago, look as though they're still begging to run.

A Dam to Tame a Dragon

Every book written on water in the West describes the Hoover Dam. More than a few open with its inspiring façade or slap it right on the cover, and why not? It was, in its day, the biggest and most important dam in the history of man. It might still be the most beautiful. Congress's appropriation of funds for its construction dramatically and directly altered the futures of seven states and set the federal government on a decades-long binge of dambuilding that would remake the nation. The Bureau of Reclamation, once little more than a ditch-digging operation, became one of the government's largest and most powerful offices. It remains the nation's leading water wholesaler and second-leading provider of hydroelectric power, and its story begins in many senses with the national landmark of the Hoover Dam. All that is great and glorious and terrible about the golden age of American dambuilding appeared, in one form or another, in what engineers first called the Boulder Canyon Project.

The Colorado River is a dragon. I don't mean the European sort of dragon, all fangs and fire, but the Chinese variety. Chinese dragons are slender and serpentine, their claws nearly vestigial, and they glide through the sky as though neutrally buoyant against the clouds' vapor. Indeed Chinese dragons are creatures of water, not fire, and, instead of vicious beasts, they are wise and generally benevolent. Symbolizing the power of water, they bring wealth and prosperity but when provoked may respond with earth-rending violence.

In its natural state, the Colorado was the most tempestuous waterway in North America. From the high crags of Wyoming and Colorado issue forth cold little streams that merge into the river's slender tail. Hundreds of miles of Rocky Mountain wilderness drain inexorably to the most direct channel and, by the time the river reaches the Utah border, it has already gathered formidable downhill momentum. On a long, meandering course through the high desert, that terrible force will tear untold tons of sand and sediment from the banks. The power that gouged the Grand Canyon into the face of the Earth never stops working. As a consequence, the Colorado's ratio of silt to water is *seven times* higher than the typical flow of the "muddy" Mississippi.[2] At times it is less a body of water than a roiling bulldozer of liquid sediment.

Falling several thousand more feet into Nevada, the river sloshes due south through southern California and northern Mexico. A broad silty delta is the golden crest on the dragon's head—indeed, Chinese mythology states that without such a growth (the *chimu*, it's called) a dragon cannot fly. At the delta's mouth stands the Gulf of California, a warm and isolated inlet on the eastern shore of Baja. Here the dragon makes its exit from the corporeal world, though for much of the last century little to no water has actually reached the delta. Every drop is drained, drunk, or diverted by thousands of claims across seven American states and one Mexican state. During that run the Colorado is dammed no fewer than fourteen times. The reservoirs standing behind those dams can store multiples of the river's annual output, pushing a serious flood nearly out of the realm of possibility. But at the dawn of the twentieth century, the Colorado was an absolute tyrant. Draining much of the Southwest's mountains and desert, it could be quite placid during dry years and then burst its banks in the span of a single stormy day—a red stampede piling up natural levees and carving new landscape like a shovel pushing sand.

Communities that sprang up on its banks might divert and enjoy its water for a season or two, but the river inevitably silted up their ditches or simply inundated their fields. California's Imperial Valley, in the desert east of San Diego, experienced a series of engineering missteps and storms in the winter of 1905 that caused the Colorado to break loose and cut itself a new gorge. The Salton Sink, an old salt flat north of the valley, attracted the loose water and became the eerily artificial puddle known as the Salton Sea. Months of labor by the increasingly frustrated and desperate landowners of the Imperial Valley brought the Colorado back to its old course, but even the finest salesmen could no longer promise anyone their farms would be safe. The same river that lent them their livelihoods also posed an existential threat. In order to turn the naturally parched West into a green, modern paradise, they'd need all of the Colorado's water without its attendant dangers.

Enter the U.S. federal government: the only entity with the resources and legal authority to organize a project spanning seven states. In 1907 President Theodore Roosevelt sent to Congress "a broad comprehensive scheme of development for all the irrigable land upon the Colorado River," girding the entire watershed with dams and reservoirs to control its legendary floods and provide cheap irrigation water for 700,000 acres of new farmland. To lend some perspective on the geographic scale of the region, the river basin itself is 246,000 square miles, or upwards of 150 million acres. Even Teddy at his most ambitious never envisioned greening more than one half of one percent of the West's immense, bone-dry wilderness.

The West as Wasteland

It's hard to explain just how dry the American West tends to be. The Mississippi and Missouri River Basins represent the limit of naturally cultivable land. Meriweather Lewis and William Clark's 1804 journey from St. Louis through the newly acquired Louisiana Purchase and up to the Pacific Northwest made clear that prospects for agriculture were grim. For all the searing natural beauty the Lewis and Clark party beheld, they found few sources of surface water. Yet the water they did find was their primary guide—they followed the Missouri's headwaters to the continental divide, then tracked the Clearwater and Snake Rivers down the western face of the Rockies until they met the immense Columbia River. The landscape itself astounded Lewis and Clark, rugged and flush with game yet almost everywhere deprived of trees. Grasses and shrubs were all that could survive more than a stone's throw from riverbanks, at least until the mountain slopes sprouted pine forests, but nobody wanted to farm or ranch on poor mountain soil. To settler Americans, no trees meant no real possibility of agriculture. General Zebulon Pike led, at Thomas Jefferson's request, an expedition through Kansas, Colorado, New Mexico, and Texas. He compared the territory to the Sahara Desert. Legendary frontiersman and trapper Jedediah Smith hiked six hundred miles across Nevada's Great Basin Desert and only encountered three faltering streams.[3]

The plains were wetter, but not enough for growing crops unless one could provide or pay for the labor required to dig irrigation canals from the nearest river. Irrigation communities sprung up here and there, often founded by congregations of faith with the strong social ties necessary to collaborate on such a difficult and labor-intensive project. Irrigation ditches and canals are backbreaking work to dig and maintain, as dry ground tends to be very hard, and any ditch will require regular dredging to keep it from silting up with river sediment. Mormon communities became excellent at irrigation farming, because their tight-knit and hierarchical culture guaranteed nobody would skip out on his ditch-digging duty. Few Western settlers could rely on support from their whole communities, so most living between the Missouri Basin and the Rockies could not reliably get enough water to raise healthy crops. Instead they grazed cattle on their land, occasionally leading the herds east to St. Louis or south to Texas for sale and returning with several years' worth of income. With the native bison populations destroyed and the Indigenous nations driven nearly to extinction, many thousands of square miles of federally owned territory became available, if not exactly desirable, ranching land.

Washington was desperate to give it away. Two generations of American leaders had used handouts of Western land to head off political conflict, and they weren't about to stop. The contradictions of a rapidly industrializing North and a southern agricultural aristocracy built on the labor of enslaved people were too great to hold forever, but few men in Congress wanted to be responsible for an existential crisis in the republic. They headed off the inevitable reckoning by encouraging western expansion, handing as much Indigenous land to new settlers as they could cultivate. For much of American history, land giveaways were the pressure valve to preserve the status quo. As the Civil War appeared increasingly inevitable, as the struggle against slavery escalated to armed conflict in "Bloody Kansas," the government was still trying to pull the same lever. In 1862 during the Civil War, Congress passed the first Homestead Act that served as a framework for all Western expansion thereafter. Instead of the large plantation-style holdings found in the South, land would be parceled out in

160-acre plots: squares half a mile on each side, also known as "quarter plots," as four fit in a one-mile-by-one-mile square. One family and plot at a time, Jefferson's nation of free yeoman farmers might finally take root.

In the East, where it rained year-round, farmers could grow anything the soil allowed and thus a quarter plot could yield a modest living. Unfortunately in the arid West, 160 acres offered only enough natural vegetation for a handful of cattle. To raise a big enough herd to turn a profit, a rancher needed several thousand acres. Government workers in Washington DC had no experience with the lands they parceled out—they were just filling in squares on their maps. As a consequence the most successful ranchers tended to be the most sophisticated, or brazen, cheaters of the homestead system. The great Western explorer John Wesley Powell, a one-armed legend of a man who led the United States' first expedition to chart the insanely dangerous Colorado River, pleaded with the feds to reorganize their vision of the West around its scarcity of water. The only way to build an American civilization out there, Powell explained, was to erect it around the principals of sharing: shared water, shared grazing land, and shared responsibilities to dig and maintain irrigation networks. Irrigation was the only way to make this work, and Powell had learned from Mormon farmers how difficult those group projects could be.

Powell also understood that the state boundaries, as drawn, made no sense in the West. A river like the Colorado ran through seven states; instead, he proposed, borders should be drawn to make each river basin its own state. Otherwise there was no way to coordinate policy, no way to establish the trust necessary to make major decisions about water—a substance that, especially in the West, was life, death, everything. Of course nobody wanted to hear Powell's objections at the time, and expansion continued with heedless abandon. Decades of hardship and drought eventually convinced the region that they needed large-scale irrigation projects, and that federal action was the only plausible avenue. The Colorado River Compact was therefore an historic attempt to design multistate policy around a single river basin.

Herbert Hoover, Dealbroker

In 1922, fifteen years after Teddy Roosevelt's grand pronouncement of a new future for the Colorado River Valley, the region's member states finally convened in Washington DC to bargain out the particulars before Secretary of Commerce Herbert Hoover. Michael Hiltzik's book *Colossus*, about the dam that would bear Hoover's name, paints the man as a buffoon, an oblivious stuffed shirt who arrived at one of the twentieth century's most contentious domestic negotiations expecting nothing of the sort. "There is little established right on the river and we have almost a clean sheet with which to begin our efforts," he announced of a watershed comprising more claims than any single register could document.[4]

The commission's delegates were savvier but voraciously competitive, each arguing for metrics and standards he expected to benefit his own state. When asked of how much irrigable land each state possessed, the delegates exaggerated their figures by nearly 50 percent more than the government's estimate. The figures they produced demanded enough water to run the entire basin dry. As Hiltzik describes, "Confronted by Hoover, the commissioners all admitted that they could not back up their figures with hard evidence. But none agreed to amend them." The session would later collapse into shouting and threats of violence between grown men arguing over a body of water more than a thousand miles away.

New Mexico hosted the next session in November 1922. This round was vastly more productive, owing to a compromise between the states to divide their interests into two blocs: the upper basin, comprised of Wyoming, Utah, New Mexico, and Colorado; and the lower, comprised of California, Nevada, and Arizona. A seven-way negotiation could thus be simplified to two major parties, who could settle other details among themselves. This simplifying notion was developed by Delph Carpenter of Colorado, but Hoover went ahead and swiped it. He later claimed the idea struck him in the dead of a sleepless night, when in truth Carpenter had been openly working on the compromise for months. Nobody present could possibly have been fooled.

The Upper/Lower compact sorted highly contentious issues into more manageable bipolar ones, but the bedrock was no easier to drill. The Colorado's highly variable flow forced the commission to work with ten-year averages. After some compromise the upper states promised an annual average of 7.5 million acre-feet to the lower states. Unfortunately for the commission, the data used to derive this figure (and many others) was incomplete. As Hiltzik writes, "Later research would establish that the 1899–1921 period was one of the wettest in the basin's history. In no ten-year period since then has the flow matched the average of 16.4 million acre-feet . . . on which the compact is based; the true historic average . . . is less than 14.7 million, insufficient to fulfill the minimum compact apportionment."[5]

The state delegations worried less about drought than competing claims that might limit their growth. In order for the lower states to secure their prospects into the future, they'd need the year-to-year flow security that only reservoirs on the river could provide. The lower states suggested damming the river in their territory, but the upper states saw that as a naked ploy by California to carve out enormous water claims for the future. The Golden State didn't have the population to use all that water themselves—not at present—but everyone knew even provisional water claims tended to become permanent in short order. In practice, once someone started using water, they got to keep it. What court or authority would parch an extant settlement for the sake of a theoretical claim upriver? Most of the greater Los Angeles region was built on a series of increasingly brazen water grabs, so the rest of the Colorado River states rightly feared any arrangement that might open the door to more power plays from the Californians.

Hoover, in his fashion, stepped on the rake. As an upper-state delegate inveighed against the proposed dam, Hoover interrupted him. "If you will allow me to become a Californian a minute instead of a Chairman," he begged of an audience who had feared exactly this, arguing, as Hiltzik paraphrases, "that the 'people of the lower river' had put so much effort into cultivating their land that they deserved to be guaranteed all the water they were already using, or more."[6] It is difficult to imagine a more boneheaded ploy. Hoover was forced to adjourn the meeting, and that night most of the Californian delegation marched out of the conference after failing to persuade Hoover to double down on their behalf. The next morning, perhaps buoyed by the absence of their Golden State colleagues, the commission passed the compact giving only cursory mention to "works . . . to control floods" on the lower river.[7] Hoover was chuffed at their success, though he had frankly done more than any other man to hinder it. In any event, the compact would only be valid once ratified by all seven states: a formidable legislative gauntlet few expected it to clear.

The Fight for Public Power

The project faced immediate opposition in the form of private electrical utilities and the politicians they'd long since bought. Private utilities saw any publicly owned generating capacity as threatening their profits. They fought any development in that direction with the same mix of dirty money and anticommunist rhetoric every threatened incumbent has deployed for a hundred years. Of course, dams are poor examples of capitalism: impacting the very terrain for miles in every direction, dams typically require the kind of legal sanction only government can provide. Once operating they generate a lot of electricity for a very long time with little overhead, which in the absence of a profit motive means they can distribute that power very cheaply. Private utilities struggle to compete against their cheaper, safer public sector equivalents and therefore attempt to defeat them using the political process. The Colorado River Compact could not meet its goals without a major expansion of public power, so utilities across the West spent lavishly opposing it.

Those companies had no closer ally than Herbert Hoover. He supported the compact's flood control and irrigation projects, but despised public power as a threat to private enterprise. For this reason he pushed for the ouster of the director of the Federal Reclamation Service (later renamed the Bureau of Reclamation), Arthur Powell Davis, whose uncle John Wesley Powell had been the first American to chart the Colorado River. Davis planned to finance the planned dams on the Colorado by selling the electricity they'd one day generate. To Hoover this was anathema, and he could not have been displeased when Interior Secretary Hubert Work dismissed Davis.

Hoover's historical reputation as an oblivious right-wing oaf strikes an odd chord against his biography. The son of a Quaker blacksmith, Hoover was orphaned by the age of ten and admitted to Stanford University as a member of its very first class in 1891, despite lacking a high school education. He excelled at mathematics and trained as an engineer, earning him the distinction of being the first Stanford-trained engineer to wreak havoc upon the United States. Following graduation he went to work as a mining engineer. Years of successful consulting and investing made him a fortune. During World War I he worked with the U.S. government to help distribute food aid to Americans and other allied civilians in Europe. Hoover earned a reputation as a model administrator and systems manager, importing his knack for industrial efficiency from the mining industry to the public sector, which earned him the appointment to secretary of commerce in 1920. You might think of him as an early modern analog of the corporate consultants who have overhauled so much of our twenty-first century civil service.

He represented the cutting edge of smart, evidence-based thinking, which in important moments he abandoned for a reflexive promarket ideology. Hoover understood that permitting the government to finance public works by selling their own power and water would hurt the profit margins of private utilities. Far worse, it would lead to a major expansion in public spending, as public works could be built without the assiduous consent of the banking industry. The fight over public power has never been about electrical prices for consumers. It is a proxy war over whether finance should control government spending, and Hoover understood very clearly which side he served. As California senator Hiram Johnson, one of Hoover's Republican rivals, once wrote, "Wherever we uncover a power trust rogue, we find Hoover."[8]

Public power won the battle by arguing on the market's terms. State legislators could not turn down the prospect of so much cheap electricity, particularly in the sparsely populated West. There was a question as to whether the Colorado River states could muster enough demand to consume what the dam at Boulder Canyon produced, until famed Los Angeles water autarch William Mulholland arrived in Washington DC to declare the rapidly expanding sprawl would be "ruined" if it could not secure additional supplies of water and electricity.[9] The city's demand might grow 20–25 percent per year, he claimed, though, like much of Mulholland's public pronouncements, this was an embellishment. An Irish-born ditch digger who rose to singlehandedly command a massive and vital public agency, Mulholland was also a shameless boaster and provocateur who held nothing on Earth in higher esteem than himself.

Much of what he told Congress was a lie. Los Angeles faced no imminent drought. Conjuring fears of desolation was one of Mulholland's favorite tactics with politicians and press. He also attested, "as an engineer," to the soundness of the proposed site. He was not in fact much of a scrutinizing force, as a dam of his design (the St. Francis Dam) collapsed on March 12, 1928, killing over 450 people. This calamity failed to derail the Boulder Canyon project, though it did end Mulholland's career. The St. Francis Dam remains one of the worst disasters in California's history and emphasized the need for tighter, federally controlled regulations on water infrastructure. The Boulder Canyon Project Act passed Congress in December 1928, even as it was re-sited downriver to Black Canyon: a superior prospect, high-walled and tapering to a narrow "gunsight" drop that limited the size of the proposed dam while increasing the size of the reservoir behind it. It was also closer to civilization, only twenty miles from a railroad line in the dusty little town of Las Vegas. Congress set the project's budget at $165 million, in a year when all other federal construction projects cost $150 million combined.[10]

The Taskmaster

Work began in late 1930, in conditions nearly impossible to imagine. Daytime temperatures regularly exceeded 120 degrees; the night air cooled to a mere one hundred degrees as famished men struggled to eat, shaking and dripping sweat on their trays. For three dollars a day (poverty wages even in the early years of the Great Depression), they cleared rubble in sweltering dark caves or swung like pendulums across sheer cliff faces to plant dynamite or donned heavy metal boots to stomp air bubbles out of freshly poured concrete blocks. During breaks they'd peel off the boots and tilt them like pitchers to pour out their sweat on the red ground. The vast majority of them had never worked on a dam before. They received no training but what they picked up on the job. Over a thousand men swarmed the Black Canyon day and night under the restive eye of the era's greatest dambuilder. Frank Crowe was a taciturn New England asshole in the classic mode, a man so settled in his convictions that he insisted his family live with him in frostbitten work camps even as they kept dying of pioneer diseases well into the twentieth century.

Crowe earned his reputation for excellence the same way any project manager does: by beating deadlines and cutting costs. There was no stress that he would not offload on the workers and no opportunity for savings he'd let slip under his beaklike nose. He was the perfect man to erect a world-historic dam on a budget in the middle of a hellish desert. Though a tyrant in every honest sense of the word, Crowe ran an organization designed to insulate himself from public accountability so as not to *appear* tyrannical. He maintained a crew of lieutenants that followed him from project to project, bringing family connections in on the business until every crew on the dam site was run by a fierce Crowe loyalist and his sons, brothers, or uncles. When labor unrest appeared, these men would smother it without Crowe even showing his dour face. As a result, most Hoover Dam workers appear to have held the big boss in high esteem.

A cruel taskmaster is born every day, but Frank Crowe was much more than that. A brilliant engineer, he spent years prior to the Hoover Dam (it was named Boulder Dam until 1947, but the later name is used here for consistency) refining a novel technique for pouring large volumes of concrete to a high degree of precision. He built steel towers to serve as anchors for a system of high-flying cables on which rode giant buckets carrying tons of concrete. The towers could also be mounted on custom-built sections of railroad track to reposition the cables at will. Rather than build interlocking portions of the dam and its massive subterranean penstocks in a steady sequence, Crowe had crews working on every component at once with the expectation that they would all finish simultaneously. This approach increases the chance of errors, but if executed well it can save a great deal of time and money. "Just-in-time" production is now the American standard across nearly every industry; Frank Crowe was decades ahead of his contemporaries.

That's not to drape the Hoover Dam's glory exclusively around his shoulders. The project demanded cutting-edge expertise from scientists and engineers around the country, whose work would lay the foundation for decades of progress to come. Concrete engineering is a delicate science with infinite variability, and the Black Canyon project posed a unique challenge. No dam had ever been built at such scale; such a quantity of concrete had never been poured. Holding up the mass of half a mountain, the dam's internal structure demanded heavier, chunkier concrete by up to a factor of six. Nobody knew exactly how this material would behave under live conditions, so three universities and two government labs spent two years testing fifteen thousand concrete samples, compiling what Michael Hiltzik calls "a hoard of data that advanced the science of concrete manufacture by a quantum leap."[11]

Crowe was indifferent to the human toll inherent to his demanding schedule and the brutal conditions of nature. Heatstroke and disease from drinking dirty river water were constant companions, but accidents became increasingly common as crews excavated the huge tunnels needed to divert the Colorado River. Men boasting zero prior experience with explosives shoved dynamite down narrow drilled holes with long sticks, rushing out just ahead of the explosion. Descriptions of onsite accidents are grim and gruesome; the lucky ones went quick. Dozens died in premature blasts, rockslides, cave-ins and vehicular accidents simply during the tunnel excavation.

The official numbers say ninety-six died during the whole project, but this is plainly an undercount, as Crowe and his employers (a conglomerate known as the Six Companies formed specifically to bid on the Hoover Dam) frequently reclassified deaths and injuries to keep them off the books. Deaths by sunstroke, dehydration, or accident, where the victims died later in the hospital as opposed to onsite, could generally be scrubbed away. Of course, workers also suffered thousands of injuries in addition to those reported. Attempts by dam workers to collectively bargain were unsuccessful. A shortsighted decision to cast out organizers from the Industrial Workers of the World sacrificed vital strategic acumen on an altar of respectability to unscrupulous bosses who gave not an inch in return.

In any case, those early Great Depression years afforded little leverage to labor. Economic devastation around the country ensured dambuilders would always have as many desperate workers as they needed. Nonetheless, the men onsite seem to have been genuinely invested in their work and excited to participate in a historic undertaking. They knew the dam would stand as a symbol of national accomplishment and took great pride in it.

Enthusiastic and generally unskilled labor allowed Crowe to reapportion workers as needed in response to unprecedented problems. When it became clear that the ring-shaped steel wall sections of the penstocks would be too large to transport by rail, Crowe had a custom foundry erected onsite to cast them. When math suggested that the incredible bulk of the dam would keep its concrete from cooling for more than a century, the Six Companies began laying thin steel pipes into the concrete blocks to pump cold water through the mass. To supply that water they built an enormous refrigeration plant on the riverbank, out in the middle of the desert. Each time Crowe and his men met a challenge, they deployed seemingly limitless resources to address it as directly as possible. Despite Crowe's obsession with costs, he reasoned doing the work fast and doing it right would benefit his employers more than cutting corners. If a solution existed within the parameters of contemporary science, those men in the Nevada desert made it reality.

Work finished in early 1936. Frank Crowe, true to his reputation, turned the Hoover Dam over to the feds more than two years earlier than planned. The historic structure required numerous repairs and renovations in the years following, but completing a project of such scale and ingenuity in such hostile conditions is extremely impressive even today. The project became the standard for all future dams, swiftly establishing itself as a major sightseeing destination for Western motorists and cementing the Bureau of Reclamation's sterling reputation for decades to come. Supplying water and electricity to Nevada, Arizona, and California in addition to nullifying the risk of flooding across thousands of cultivated acres, the Hoover Dam also became a model for water infrastructure as the engine of regional economic development.

Grand Coulee

With Franklin Roosevelt in the White House and the Hoover Dam standing as an avatar of American potential, money for infrastructure was easier to come by. One of the administration's highest priorities, a central column of the New Deal, was economic investment in poor and rural America. They did this not only because those areas suffered terribly in the Depression, not just because they thought it was the right thing to do, but because it placed money and economic leverage in the hands of working people who could contribute to an enduring electoral majority. Indeed, much of the rural South and West responded by voting Democratic for decades. Without Roosevelt's commitment to rural electrification, it is entirely possible large swaths of our country would still not enjoy consistent access to electricity. The city of Flint, Michigan, has been without safe drinking water since at least 2014 with no end in sight, so obviously a program of electrification like that of the New Deal would be impossible. But in the 1930s when the federal government decided an objective was important, it devoted real resources toward completing it—even if Congress had to be led by the nose.

Trickling down as winter runoff from western Canada's Purcell mountain range, the Columbia River gathers strength from hundreds of tributaries on its looping southward course for the U.S. border. By average flow it is the fourth-largest river in the United States—the mercurial Colorado, for comparison, ranks thirty-eighth. The Columbia is an awe-inspiring body of water, so broad and deep that the currents and eddies recede from view and the river seems more like a limb of the Pacific Ocean. Prior to Frank Crowe's triumph in Black Canyon, no one seriously contemplated trying to dam the whole Columbia. Major General George Goethals, the man who built the Panama Canal, advised the government to build a small irrigation diversion and call it a day.

By 1933, Franklin D. Roosevelt was desperate to put people to work, and the Hoover Dam provided a good model for large public projects in the name of rural development. If the mighty Columbia were to be dammed, the Bureau of Reclamation determined, it was best done in central Washington, in a region of bizarre topography known as the "channeled scablands." An ancient, tectonically induced flood (described vividly in Marc Reisner's classic book *Cadillac Desert*) blasted away most of the region's soil, exposing the bedrock like black volcanic bones, spontaneously carving wide canyons in the stone that are described today with the French-derived term *coulees*.[12] One of the largest, a sixty-mile gash called the Grand Coulee, piped the river steeply downhill inside high granite walls that would easily contain a future reservoir. The area was in many ways perfect for a tall hydropower dam.

The New Dealers faced two problems: first, no real market existed for the power they proposed to generate. Only three million people lived in the entire Pacific Northwest, the vast majority in farming and logging communities with no access to electricity. Better, the Army Corps of Engineers argued, to build a low dam for irrigation and flood control while saving a pile of government dollars. The second problem was Congress, which, despite the Democratic majority, had little appetite for outsized spending in the nation's most remote corner.

They agreed with the corps and allocated funds for a low dam, to be built by the Bureau of Reclamation.

Undaunted by this trifling matter of federal law, the bureau started work in the eerie, monochrome rockscape of central Washington. It just so happened that the foundation they laid was one appropriate for a high dam. As Reisner memorably wrote: "After $63 million had been spent building a foundation for it, however, a low dam was out of the question; at the very least, it wouldn't have made much sense. The bureau had presented Congress with a fait accompli in the form of a gigantic foundation designed to support a gravity dam 550 feet tall. To build a two-hundred-foot dam on it would have been like mounting a Honda body on the chassis of a truck."[13]

Interior and Reclamation decided they would rather beg forgiveness than ask permission. And their gambit worked. Congress gave them the rest of the money, and the product was the biggest dam on Earth. Every component and onsite resource was also the largest in the world, from the pumps to the penstocks to the concrete mixing plant. The cable-and-bucket pouring system, the coolant pipes to help the concrete dry, and every other lesson and practice established at the Hoover Dam (and California's Shasta Dam, another Frank Crowe production) got deployed in Washington state at even greater scale. As Crowe had, the Grand Coulee engineers solved problems with audaciously improvised solutions. When a mudslide threatened to bury the Grand Coulee Dam's half-built foundation, government workers scoured the northwest for any refrigeration unit they could buy, hauled them back to the scablands and used them to freeze the slide in place while construction proceeded.

In 1941, the federal government paid Woody Guthrie to write a song commemorating the world's newest wonder. Early the year following, the Grand Coulee Dam's gargantuan generators swung into action for the first time. Washington, Idaho, and Oregon suddenly had a massive electrical surplus, an inundation of free energy as relentless as the queen Columbia. But between the release of Woody's stirring folk tune and the dam opening, an incredibly momentous event occurred that would change the American West forever: the attack on Pearl Harbor, December 7, 1941.

As the Pacific Fleet burned in shallow water the color of new jade it was clear to all that the United States would go to war. Equally clear was the new shape that war would take compared to those prior. The shattered, helpless behemoths of Battleship Row demonstrated the future of naval warfare lay in aviation: to prevail, the United States would need many thousands of new airplanes.

Aluminum is the most important material in the production of combat aircraft, nearly as strong as steel at a fraction of the weight. Aluminum's metallic form does not occur in nature; it can only be created via electrolysis, using electrical current to separate the pure metal from a solution of aluminum sulfide (this solution is smelted from bauxite, a red-colored sedimentary rock). Making aluminum consumes a lot of energy, hence its relatively high price, but, in the 1940s, it was one of the most expensive manufacturing commodities on Earth. A single company, Alcoa, controlled the nation's supply of aluminum in an outright monopoly. Despite the pleas of the U.S. government, Alcoa refused to increase production lest it lower their prices and hurt their profit margins. In the end, the Roosevelt administration decided to produce the aluminum themselves.

They wanted a site on the West Coast, where these new weapons could be loaded directly onto ships and rushed into the Pacific Theater. Combat aircraft demanded specialized factories and machine shops—entire chains of supply—that would have to be built from scratch, so federal planners preferred a region with little preexisting industry. Lastly, they knew their new production center needed access to an incredible amount of electricity. The Pacific Northwest, from Portland to Seattle, was a green quilt of farmland and logging claims with barely any industry that did not involve processing lumber. It seemed a blank canvas, and, thanks to the freshly bridled Columbia River, it just so happened to possess the world's largest electrical surplus. The Grand Coulee Dam, at its conception such an eye-gouging boondoggle that the Roosevelt Administration had to dupe it past Congress, found its market just in the nick of time.

In the space of a few months, bauxite smelters and aircraft factories were going up all over the Northwest. High casualties in Europe and the Pacific kept those plants consuming all the material the government could supply through the end of the war. Rosie the Riveter was America's World War II industrial icon, but in reality the single biggest driver of the war effort was the Columbia River. Grand Coulee Dam, and its downriver cousins like the Bonneville Dam, gave the United States an advantage in electricity, aluminum, and aircraft the Axis powers couldn't hope to match. When the defense industry needed more capacity, turbines awaiting installation at Shasta Dam were rushed up to Grand Coulee. Unfortunately these turbines ran the opposite direction from the Coulee models, so the Bureau of Reclamation installed them in jerry-rigged pits that piped in water from the appropriate side. After the war, they had to invent special extracting machines to pull these turbines out of the ground because nobody had ever tried to do such a thing before. Albert Williams, who wrote a 1950 tract on Western hydropower, estimated "more than half the planes in the American Air Forces were built with Coulee power alone."[14]

Philip Nalder, a Washington State College engineer who worked on the Grand Coulee Dam during construction and managed the dam from 1950 to 1960, recalled prior to his passing in 1998 that the dam's generators were rated for a maximum output of 105,000 kilowatts each, yet ran at 125,000 kilowatts twenty-four hours a day for the duration of the war without ever failing. Like the black iron dynamos of the Folsom Powerhouse, these were *magnificent* machines:

> We would shut one down only when it was absolutely necessary. You'd stand there in the powerhouse and feel that low vibration . . . and you'd feel certain that they were going to burn themselves up. And you'd think that maybe the course of history depended on these damned things. But they never overheated, so we just ran them and ran them. God knows, they were beautifully made. By the end of the war, at Grand Coulee, we were generating 2,138,000 kilowatts of electricity. We were the biggest single source of electricity in the world. The Germans and the Japanese didn't have anything nearly that big. Imagine what it would have been like without Grand Coulee, Hoover, Shasta, and Bonneville. At the time, they were ranked first, second, third, and fourth in the world. We had so much power at Grand Coulee that we could afford to use two generators just to run Hanford.[15]

The Hanford Reservation was a secret research base used by the Manhattan Project to produce plutonium for nuclear weapons. Plutonium, even more so than uranium, is incredibly energy-expensive to refine and enrich. The immense electrical infrastructure established in the Northwest by the Grand Coulee and Bonneville dams, among others, was instrumental in securing American military supremacy throughout the twentieth century. For decades Boeing was the region's single largest employer, and to this day it is impossible to imagine the Seattle area or much of the Pacific Northwest without the aviation industry, without dense clusters of hangars, runways, and submarine bases. The broad middle-class prosperity cemented by those industries allowed the children of Washington state to grow up as tech moguls, as artists, as musicians. Certainly Boeing happened because of the Grand Coulee, but so too did Microsoft. So too did Soundgarden, Nirvana, and Alice in Chains. Dams, and the economic systems they come to anchor, have shaped everything we know about this country. Once World War II was won and the United States took its place for the first time as a global superpower, the nation found itself more prosperous than ever before. We could have done anything with it. What we decided to do, among other priorities, was to dam nearly every river on the continent.

3

THE DROWNED EMPIRE

If aridity and drought are afflictions endemic to the American West, most of the South has the opposite problem. Rain falls throughout the year, keeping streams and rivers high enough that one good surge, whether it comes from spring melt or autumn storms, may trigger a flood. For all of preindustrial history, well into the twentieth century, much of the South experienced heavy flooding on a regular basis. This was especially true for communities along the Mississippi River, positioned at the mouth of a funnel draining much of the continent. In the spring of 1927 the Mississippi flooded so badly that the resulting calamity changed American society forever.

From the 1600s through the end of the Civil War, the southern states ran an agricultural economy powered by the labor of millions of enslaved people. Harvesting cotton demanded miserable toil in the sultry and often malarial conditions of the region. It is a thirsty plant, demanding ample water and a frost-free growing season, so the warm and rainy South provided an ideal environment. Plantations adjoining major rivers were especially profitable, as river water could be diverted to the fields in dry years, and products could be easily trafficked by barge. So it should come as no surprise that the Mississippi Delta, where the nation's biggest waterway widens into a two-hundred-mile-wide complex of marshes, stood as one of the world's cotton capitals. For millennia the delta served as a floodplain, nutrients concentrating in its soil with every season, and those growing conditions made the region commercially irresistible. Around every finger of water sprung up cotton fields worked by enslaved people who lived in ramshackle conditions on the same floodplain. When the water swelled and rose, flooding the fields as well as their homes, those people were made to work digging outlet ditches to drain them. As soon as the water receded, they were made to shovel earth by the ton into levees that they hoped might hold back the next flood.

The floods kept coming, higher and fiercer with each passing decade. For all of the nineteenth century the United States pursued a "levees-only" strategy along its major rivers. Early engineers saw how running water dug deeper channels and concluded that building levees would deepen rivers, attacking the flooding problem from both directions through a lower bed and higher banks. Yet as the nation's river systems grew more developed and populated, as more and more levees went up to safeguard their property, every major flooding event was worse than its predecessor. The higher the levees mounted, the higher the floods surged. How could this be?

Ellet and Humphreys

Charles Ellet Jr. lived as though he were the protagonist of an aspirational novel. Born on New Year's Day, 1810, to a middle-class family near Philadelphia, he was working as an assistant canal clerk by age seventeen and exhibited such brilliance that at twenty he had taught himself French and persuaded the Marquis de Lafayette to grant him admittance to the best engineering school in the world: the École des Ponts et Chaussées in Paris. He returned as one of the only European-trained engineers in the United States and found immediate work building bridges. Ellet was by all accounts handsome, capable, athletic, brimming with confidence to the point of arrogance, and possessed of a real panache. He became the first man to traverse the Niagara Falls gorge when he strung a wire and pulleys over it, hung a basket from the wire and pulled himself across over 240 feet of empty air. As historian John Barry describes, "Then he built a catwalk of planks without guardrails, and was the first to cross it too, driving a horse and carriage, standing up like a charioteer, speeding and swaying, and transforming himself into a legend."[1] An illustrious career followed, and soon Ellet was one of the top civil engineers in the country.

In 1850 President Millard Fillmore assigned Charles Ellet to perform an engineering survey of the lower Mississippi River, from its confluence with the Ohio River at Cairo, Illinois, to the Gulf of Mexico. It was the most prestigious, desirable post a man in his profession could have received. There was just one problem: the job had already been promised to another man who was Ellet's perfect counterpart in nearly every way: an ambitious Army officer named Andrew Humphreys.

Humphreys was born in November 1810 just a few miles south of Ellet in Philadelphia proper. Humphreys grew up a loutish teen given to misbehavior, so his wealthy parents did as many of their station would and sent him to West Point, which had been founded as an engineering school. The experience profoundly affected Humphreys, teaching him discipline while nurturing the megalomaniacal tendencies that would appear throughout his life. Humphreys considered himself destined for greatness and glory, though at least he was willing to work for it. He joined the Army Corps of Engineers and found little success, but, through ambition and social maneuvering, he made more than a few friends in high places. Finally, at the end of his thirties, Humphreys's connections paid off, and the Secretary of War assigned him to the Mississippi River report. He had finally found, he claimed, "the work of my life." But the civilian side of the government, the civil service and members of Congress, had no interest in leaving this important task to the military. They lobbied the president hard to hire Ellet, and Fillmore agreed to split the project by having both men write separate reports. They would split the project's $50,000 budget.

Obviously Humphreys was enraged—he felt the rug had been pulled from under him, his greatest opportunity for glory watered down by this *celebrity*. There was a competition underway, and he was determined to win. Not only would his report be the biggest, best, most detailed report in all of American history, he would get it out before Ellet. Humphreys worked obsessively, spending long weeks on the river in murderous heat, always dressed to the nines in a starched uniform. His behavior grew erratic; he stopped writing to his wife. He recorded data with such granular detail that he would eat samples of riverbed clay, documenting their texture and flavor.

Charles Ellet had already made his name several times over and wasn't about to sweat this gig. Surely he knew of Humphreys, but the two never met, and Ellet never seems to have mentioned him. Ellet traveled the Mississippi, had his assistants perform surveys and compared them against the extensive work he'd already done surveying the Ohio River. But he also brought his wife and children along to New Orleans and made a time of it, taking in P. T. Barnum's circus and a concert by Jenny Lind, the world-renowned "Swedish Nightingale." Ellet submitted his report to the federal government in October 1851. At that moment Andrew Humphreys was in Philadelphia, bedridden due to a nervous breakdown.[2]

Ellet's report was a seismic event in the engineering community. He hadn't collected much data but demonstrated a phenomenal grasp of the river's properties and the ability to convert those observations into actionable conclusions. He announced that several centuries of accumulated European knowledge on river flow and sedimentation were flawed—rivers did not, as previously believed, carry their maximum sediment load at all times. As a consequence, he determined that American engineers had been wrong about levees. Since the most valuable farmland lay in alluvial basins like the Mississippi Delta, the levees built to protect crops were denying the river its natural flood plains. The faster river flow dug out a deeper channel at first, but this effect was subject to steeply diminishing returns as the sediment load dropped. When the river expanded, its water needed somewhere to go, so it sought out its natural floodplain: the farmland behind the levees. Thus, Ellet declared, the levees meant to prevent floods were actually making them worse. He called them "a delusive hope, and most dangerous to indulge, because it encourages a false security," and continued, with deliberate italics, "The water is supplied by nature, but its *height* is increased by man. *This cause is the extension of the levees.*"[3] Instead of the levees-only policy, Ellet proposed an enormous multistate network of dams, reservoirs, and outlets to control and redirect floodwater. Levees would stand as the last line of defense.

Humphreys spent the next several years wracked with psychic agony, working on railroad projects for the Army, insanely jealous of Ellet and determined to finish his report even as the survey moved out of their offices and sent their findings to storage. In 1857 he resumed working on the Mississippi project and, in late 1861, he finally published his report. Unfortunately, nobody had much time to review it, as the Civil War had begun months before. Humphreys, still a commissioned army officer, received the combat assignment he had always craved.

In the heat of battle Andrew Humphreys was God's perfect idiot. He immediately earned a reputation for personal courage, advancing past stupidity into something approaching psychosis, yet nothing could touch him. Any officer who had gone to West Point and didn't get himself killed rocketed up the ranks. In eight months Humphreys was a brigadier general. At Fredericksburg, amid the carnage of an utterly futile charge ordered by General Ambrose Burnside, Humphreys lost a thousand men in twenty minutes—one fifth of his command, including five of seven junior officers. He didn't care. In fact, he felt he touched the sublime: "The charge of my division is described by . . . some general officers [as] the grandest sight they ever saw, and that as I led the charge and bared my head, raising my right arm to heaven, the setting sun shining full on my face gave me the aspect of an inspired being. . . . I felt like a young girl of sixteen at her first ball. . . . I felt more like a god than a man. I now understand what Charles XII meant when he said, 'Let the whistling of bullets hereafter be my music.'"[4]

Over the next year Humphreys's division suffered hideous casualties. They got demolished in nearly every engagement, as Humphreys himself led every charge and somehow emerged clean as a whistle. His men knew him as an insufferable dandy, changing his uniform the moment it became anything less than immaculate, "so boyish and peppery that I continually wanted to laugh in his face."[5] But no bullet could catch him, and he fought with the maniacal courage of a man who knew himself untouchable. He seemed not to understand why the same rules did not apply to his men. Still he thought himself brilliant, and who could gainsay the man? After getting beaten at Gettysburg so badly a distant scout called Humphreys's division "the crater of a volcano of destruction," he was promoted to major general and gave a farewell speech entirely about himself.[6] In his new position on the staff of General Gordon Meade (himself an engineer by training), Humphreys was at least no longer in a position to get others killed.

Humphreys emerged from the war a powerful figure in the Army and one of the world's foremost civil engineers, for his report on the Mississippi had spent the war years circulating amid the global scientific elite. It was thorough to the point of being exhaustive, replete with diagrams and data whereas Charles Ellet had largely produced a work of theory. That contrast served as the basis for the assault Humphreys would launch on Ellet's reputation over the coming years. It wasn't enough to call Ellet's work spotty or slapdash or even lazy. Humphreys declared that in three centuries of hydrological surveys conducted across two continents, Ellet's recommendations for the Mississippi were "the worst ever suggested."[7]

The Drowned Empire

Humphreys had to confront one very serious problem: for all the detail-oriented work he had poured into his report, the charts and appendices it boasted that Ellet's work did not, it nonetheless pointed to the same conclusions. He found, as Ellet had, that the European models of sediment delivery were fatally flawed, and, as a consequence, the Mississippi basin needed outlets and reservoirs to handle its seasonal floods. Had Humphreys been motivated by anything but his own glory, he might have understood that confirming a colleague's earlier supposition with further data is the core work of science. Instead he found the notion intolerable. If his report pointed toward the conclusion that Ellet had been right, then the conclusion would have to change. Humphreys cooked up a few disingenuous objections, slid them in at the end of his report like runny eggs, and the engineering community slurped them down hardly knowing any better. This man was the world's foremost expert, after all.

Charles Ellet couldn't object, because he was dead. Though no trained soldier, Ellet was a patriot and offered his services to the Union. He conceived and proposed a fleet of steam-powered rivercraft that would act as ramming vessels to help the Union maintain their blockades of the Confederacy's Mississippi River ports. The navy was skeptical, but Secretary of War Edwin Stanton gave Ellet the personal dispensation to pursue his project with the rank of Colonel of Engineers. In just two months he had converted nine of the fastest steamboats on the Mississippi into sleek, low-profile warships featuring deadly armored spurs that jutted from their bows below the water line. Ellet's personal panache extended to his wartime efforts: he named his flagship USS *Queen of the West* and affixed a large ornamental letter Q on the angled armor plating between her smokestacks. His flotilla joined the Union's Mississippi River Squadron on May 25, 1862, and steamed downriver to Memphis, Tennessee.

Once moored in the muddy river in sight of Memphis, Ellet

was deliberately left out of the battle planning. Navy captain Charles Davis placed no faith in the ram ships. Undeterred, Ellet seized the initiative once battle was joined. *Queen of the West* surged ahead of the pack, joined by Ellet's brother Alfred at the helm of the ram ship USS *Monarch*. In minutes they pierced the Confederate battle line, forcing their guns down while the Union gunships closed from behind. *Queen of the West* and *Monarch* both slammed at flank speed into the Confederate flagship CSS *Colonel Lovell*, a "cottonclad" steamship protected from enemy fire by literal bales of cotton lashed to the gunwales. *Queen of the West's* spur struck below this makeshift armor, harpooning the larger warship in the belly. The brothers Ellet separated their ships from the stricken *Colonel Lovell*, already doomed and sinking, as CSS *Sumter* and CSS *General Beauregard* opened merciless fire at close range. Musket balls perforated the hull. Cannon blasts sheared off one of the *Queen's* paddlewheels and sent her wheeling toward the riverbank while Alfred Ellet covered his brother's retreat by sending *Monarch* into the side of *General Beauregard*. The *Beauregard* took on water and her commander plowed her into the riverbank beside the *Queen*. Ellet organized his grounded ship's crew into a boarding party and captured the enemy craft as the Union Navy, and the remaining rams proceeded to sink seven of eight Confederate ships.

The Battle of Memphis marked the end of Confederate military efforts on the Mississippi. From that day on, the Union held near total control over the river. The Union forces suffered just one casualty in the battle: Charles Ellet, felled by a sharpshooter's bullet to the knee. Though doctors recommended amputating his lower leg, Ellet declined their advice. He gambled wrong, dying on June 21, 1862, after his wound became infected. Ellet was buried as a hero in Philadelphia. It is safe to assume he went to his reward never thinking even once of Andrew Humphreys.

Humphreys, sour apple that he was, could never let anything go. He spent the rest of his God-gifted longevity terrorizing his subordinates at the Army Corps of Engineers and making his family miserable (his son described a father motivated entirely by spite and negativity, who attempted to instill the same values in his children). His crusade against the conclusions of Charles Ellet, combined with his influence with the corps, imposed decades of wrong-headed and ultimately disastrous river management policy upon the people of the United States. Levees-only would remain the law of the land until such time as a truly Biblical disaster might force the government's hand.

Delta Blues

The Union Army and federal government put an end to de jure slavery at the Civil War's end, and, at first, conditions for the typical Black Southerner improved. Freedmen who had been forbidden from attending school, owning property, or standing for elections began to do all of these immediately and with distinction. But this rapid progress only occurred because the North held the South under military occupation, and that state of affairs could not persist forever. Southern leaders bellowed for the freedom to which they had become accustomed: license to subjugate millions of Black Americans and steal anything they might possess. President Abraham Lincoln was wary of ending the occupation without first securing full suffrage and property rights for the freedmen. Whether a genuine multiracial democracy could have been established in the South, with all the antebellum power centers shattered and a powerful president committed to an emancipationist vision, is one of the most consequential and painful questions in American history. It will forever stand unanswered, because Confederate sympathizers assassinated Lincoln in April 1865. Enter the vice president, reactionary former slaveholder Andrew Johnson, a weak and widely despised figure who swiftly capitulated power back to the traitorous Southern aristocracy. Seeking

peace and comity in the face of a foe with no authentic interest in either, Johnson sold out both the Northerners who'd
suffered in the war for emancipation and every Black American living in the South.

In 1890, following long campaigns of terror and murder, ex-
Confederate white supremacists used their control of the Mississippi state legislature to write a new state constitution that
largely barred freedmen from exercising their political rights.
From there Black Mississippians experienced a swift descent
back to conditions of de facto, as opposed to de jure, slavery.
Legally they were free but they were ruled by the same violent and vicious racists who had enslaved them. Black farmers worked the same land they had while enslaved, but now
they paid their former masters for the privilege: sharecropping meant every farmer owed a portion of his harvest to the
landowners, in addition to moneys due for tools, clothing,
seed, water, and the rest. This exploitative system meant that
only the luckiest, cleverest, or hardest-working sharecroppers
ended their seasons above water, metaphorically speaking.

Meanwhile, the literal water loomed. The Mississippi applied
constant pressure to a byzantine system of levees, canals,
and outlets maintained with very little coordination or planning. The works themselves were made from piled earth and
lashed lumber; nobody really expected them to last more than
a few years. Inevitably the water would find its way through,
driven by runoff or storms or merely a slow seeping. What
seemed a solid bulwark might cave in without much warning,
a whole section of the levee liquefying into muck as ten thousand gallons of water burst through the crevasse. When this
happened—and it always did, sooner or later—landowners
and police formed posses and fanned through the area, sweeping up any able-bodied Black people to repair the levees. Often
this coercion happened down the barrel of a shotgun. If someone refused, he might be beaten and dragged to the worksite,
or he might be executed on the spot. The necessary repairs

demanded weeks of backbreaking labor, conducted in malarial conditions from tent cities atop the levees themselves (often the only land still above water). Once the work was finished the sharecroppers were released to tend to their own holdings, which by that point might not be salvageable. Even the most industrious, fortunate sharecropper could spend years scratching out a living, overcoming the long odds the system stacked against him, only to see it all destroyed by nature's caprice.

The spring of 1927 was the wettest on record in the Mississippi basin. The lower river saw rain nearly every day in March. On April 8 the storms accelerated. An observer in the delta city of Greenville wrote, "I have seldom seen a more incessant and heavy downpour until the present moment. I have observed that the river is high and it is always raining . . . we have heavy showers and torrential downpours almost every day and night. . . . The water is now at the top of the levee."[8] Up and down the Mississippi basin this growing force found its way past levees—even Pittsburgh, on the distant Ohio River, caught eight feet of water in its streets. When those tributaries drained, they did so southward into the Mississippi's main channel. Every gallon of water in all those rivers had to come through the delta, and the resulting flood was one of the largest in recorded history. John M. Barry's book *Rising Tide* describes every aspect of the flood and recovery in wonderful prose, but this comparison is most illuminating: in 1993, when the Mississippi River caused the most expensive and damaging flood in the history of the United States, the river at its peak carried one million cubic feet of water per second (a cubic foot is about 7.5 gallons). In the 1927 flood, three million cubic feet deluged the Mississippi Delta every second.[9]

The fatally flawed systems of flood protection the United States had erected along the lower Mississippi were bombs waiting to go off, and the first discharge would occur at Mounds Landing on the morning of April 21. A small leak appeared in the base of the levee there, grew at a prodigious rate despite everything the locals threw at it, and finally blew a hole clean through the manmade hillside. Nobody made any attempt to account for the workers lost in the crevasse's violent birth; they were Black, after all. Those present seem to agree the number was at least one hundred souls. The National Guard noted only that no guardsmen were killed.[10]

The crevasse at Mounds Landing was, and remains, the largest in Mississippi River history. It loosed enough water to inundate five thousand square miles up to twenty feet deep. Nearly two hundred thousand people rushed to high ground where available, but in that flat region many could only climb to their roofs. Savvy families knew to throw open their windows and doors as they fled, as an open building offered less resistance to the flood and might not be swept off its foundation. At Mounds Landing the force of the water was so great that it gouged the earth like a shovel and left behind a lake that remains to this day. Boats attempting to flee downriver found themselves driven backwards, *upriver*, by the crevasse's deadly suction.

News of the breach flew south over crackling telegraph wires: "Levee broke at ferry landing Mounds Mississippi eight A.M. Crevasse will overflow entire Mississippi Delta."[11] Across the lower Mississippi, church bells began to peal. Locomotive engineers leaned on their steam whistles, and all who heard the commotion knew the river came for them. They could only pray and use what little time remained to prepare. Those with houses on high ground drew bathtubs full of drinking water; the sharecroppers, on the lowest and, therefore, most valuable farmland abandoned everything. Paradoxically, safety often meant climbing the levees, moving *toward* the river, rather than away from it.

The delta became an ocean—cruel the way oceans are, swept by cold wind, wracked by whitecaps. Tens of thousands of people sheltered atop buildings whose foundations were being eaten away. Wave action could wrap around a home, cycling like a drill bit, coring out the soil underneath until the resulting sinkhole devoured the structure and everyone who had trusted it with their lives. For the better part of a century the United States had gambled exclusively on the levees to safeguard its most valuable cropland. A few hard months of rain had exposed it all as folly, and now more than half a million people were refugees in their own country.

Anyone with a working boat put it into immediate service as others rushed to buy or engineer their own watercraft. Lumber workers lashed together simple rowboats. Bootleggers sent their quickest runners, lifted them over the swamped levees, and ventured out onto the inland sea for rescue missions. No formal authorities existed to undertake the work, so people made do themselves. In days the rescuers started to coordinate with steamboat operators, using them as mother ships where refugees could be safely deposited. Every day on the new sea was hard, desperate, and dangerous. Waves could capsize small craft, or submerged fenceposts might pierce their keels. Worse yet, desperate people might overload the boats, swamping them all—many operators carried guns for just this reason, and a few were forced to use them. Will Percy, a lawyer in Greenville and son of Mississippi senator LeRoy Percy, wrote: "For thirty-six hours the Delta was in turmoil, in movement, in terror. Then the waters covered everything, the turmoil ceased, and a great quiet settled down. . . . Over everything was silence, deadlier because of the strange cold sound of the currents gnawing at foundations, hissing against walls, clawing over obstacles."[12]

The river's action mirrored the human tragedy unfolding on high ground. Nearly everything the delta's citizens relied on to feed and shelter themselves had been destroyed in a day. A terrible hunger set in across the stranded population. To lead the relief effort on the Lower Mississippi, President Calvin Coolidge fingered one of his least favorite lieutenants, a man he declared "has offered me unsolicited advice for six years, all of it bad": Commerce Secretary Herbert Hoover.[13]

During World War I Hoover had served as head of the U.S. Food Administration, responsible for rationing and distributing food domestically, and prior to that post he had administered food aid to civilians in devastated Belgium. Yet, in the words of John Barry, "Herbert Hoover was a brilliant fool. He was brilliant in the way his mind could seize and grapple with a problem. . . . He was a fool because he deceived himself. Although considering himself as objective and analytical as science itself, in reality he rejected evidence and truth that did not conform to his biases, and he fooled himself about what those biases were."[14] Chief among those biases was seeing other people as components, replaceable cogs unburdened by independent thought or initiative. Hoover believed in applying the principles of engineering to politics and management. That meant top-down directives, backed to the hilt by an aggressive public relations machine. Hoover looked at the stricken Lower Mississippi and saw a workshop where his ideas could be implemented with minimal political resistance.

He set about coordinating the distribution of supplies. Day and night, telegrams poured into Hoover's field office in Memphis, Tennessee, detailing what various locales needed to get by. A Red Cross purchasing agent stood by to take bids from a platoon of shouting sales representatives. From the beginning, Hoover centralized policy-making power in his office while delegating distribution to county-level Red Cross organizations. This arrangement served all parties: it funneled a huge amount of money to the Red Cross, it cut administrative costs and—best of all—allowed Hoover and his lieutenants to avoid blame for any errors. If something went wrong, it would be "squarely on the local community and not the national organization."[15]

Weeks following the breach at Mounds Landing, the Mississippi's floodwaters had nowhere to go. The soil beneath had long reached total saturation, and the region's topography left no more open plains to inundate. But water is always on the move, always seeking the slightest gradient, and, on the lower Mississippi, gravity pulls toward the Gulf of Mexico. Water that had burst through levees to spread west and east of the river now began to flow south, seeking to return to the main channel. At first those downriver of the floods in Mississippi and Louisiana hoped the devastation around Mounds Landing would spare their lands and homes. Now it became clear they had bought only time. On April 30, Herbert Hoover used one of America's very first national radio broadcasts to describe the struggle: "It is a great battle against the oncoming rush, and in every home behind the battle line there is apprehension and anxiety. Every night's reading of the water gauges is telegraphed to the remotest parts of those states—a sort of communiqué of the progress of the impending, threatening invasion of an enemy. It is a great battle which the engineers are directing."[16]

But against the might of the river all men were powerless. When the floods rejoined the Mississippi's main channel, they placed untenable pressure on the levees at those points. New ruptures appeared in the armor men had built around the river. The crevasse that opened at Cabin Teele, Louisiana, established an inland sea 125 miles wide. The only dry land was the tops of the unbroken levees, treetops the only spots of green in an ocean of muddy brown. In Melville, Louisiana, surges from two separate crevasses met with "the sound of a thousand freight trains" and ripped apart a steel railroad bridge, burying the town in silt.[17]

Early May saw more rain, and circumstances worsened. The Red Cross did its best to evacuate communities ahead of the water, ushering their people into tent cities atop the most secure levees. These concentration camps (the Red Cross's term) were maintained at Red Cross expense, though able-bodied individuals were expected to contribute labor to maintaining the levee. It was essentially a feudal arrangement, but in many cases the best the mostly Black sharecroppers could have expected. Some early flood victims saw waters recede sufficiently to plant the season's cotton, only to lose it all when a June swell plowed past the ruined levees. Hoover's teams efficiently scooped up those who needed rescue, moved them to temporary shelter and tended to their immediate needs. Yet three months after the crisis began, high waters blocked any attempt at reconstruction. The living conditions on the lower Mississippi deteriorated at a steady pace.

The sharecroppers of the delta had already been some of the poorest people in the nation, and now the worst natural disaster in American history was rolling right over them. John Barry writes, "Tents had finally arrived for shelter and the weather had turned warm, but the tents were not floored and cots had not arrived, so refugees still slept on the wet ground. There were no utensils or mess hall. Blacks had to eat with their fingers, standing or squatting on their haunches like animals. Beyond the line of tents . . . were thousands of livestock. The stench was unbearable."[18] National guardsmen kept watch with their rifles and engaged in every abuse one might imagine, from beatings and robbery to sexual assault. Will Percy, placed in charge of the flood response in the city of Greenville, knew the situation was intolerable and getting worse. He decided to evacuate as many Black people as possible from the levees, loading them on steamships bound for ports on the upper Mississippi.

It did not happen. Will Percy faced a formidable obstacle in his father LeRoy Percy, who backed the delta's landowners in their furious opposition to the plan and scuttled it before a single ship could be loaded. The landlords knew their tenants' lives had just been completely destroyed. They had nothing anymore, owned nothing but debts, were camping atop soggy levees with their livestock while being forced to work for nothing. They'd jump at the chance to resettle in the North—and they'd never come back. The workforce the bosses made their riches exploiting would depart in droves. Under no circumstances could the sharecroppers be allowed to leave. LeRoy Percy told reporters that the delta faced "the devastation and abandonment of a once-proud county. The stake is human lives and an empire."[19]

The situation in Greenville grew so dire that rumblings appeared in the press, especially outlets run by Black Americans or progressive activists. The Democratic and Republican parties of 1927 were far less homogenous ideologically than they are in the twenty-first century—both counted progressives, centrists, and conservatives among their ranks. Republicans, the party of Abraham Lincoln, enjoyed the support of most Black voters, and news of the shameful conditions at Greenville spread quickly in Black communities nationwide.

Herbert Hoover intended to run for President in 1928 and understood Black Republicans held the key to his presidential ambitions, as much of the party's right-wing establishment disdained him. Humanitarian conditions on the lower Mississippi, particularly in the camps at Greenville and Vicksburg, quickly won Hoover's attention. At his command stood state and federal governments, the Red Cross and the national guard. He wielded more unilateral power over that rich-yet-poor region than any man had since Lincoln following the Civil War.

Do not take this to mean that the sharecroppers had found a friend in Hoover. His earnest interventions into the delta's economy fared as well as one would expect for a mining engineer dabbling in agriculture. Attempting to diversify the region's crops, he ordered the Red Cross to purchase enough soybeans to plant four hundred thousand acres, even though advisors warned him the summer was far too late to plant soybeans. Hoover, ardent believer in markets, convinced himself the delta's biggest obstacle was lack of access to credit rather than the persistent and deliberate poverty in which so many people languished. He embarked on a convoluted scheme to raise money for loans from the bankers and business leaders of the South, only to be bitterly disappointed when they spurned his entreaties. He struggled to understand why they took no responsibility for the people over whom they ruled. In a meeting in Memphis, he asked local elites to pony up $200,000. When they refused, he exploded into uncharacteristic rage, bellowing curses at their fecklessness. Then he issued an ultimatum. Everyone in the room knew tens of thousands of Black laborers were at that very moment camped atop levees in flooded Memphis. If the money wasn't forthcoming in three hours, Hoover promised, "I'll start sending your n--s north, starting tonight."[20]

The bankers caved. Hoover had found their pressure point. His fundraising fortunes improved after that confrontation, though the $13 million in new credit he secured was almost nothing if divided among the flood's hundreds of thousands of victims. Even where money was available, most sharecroppers had already mortgaged everything on the year's planting, now ruined. They had no collateral with which to obtain loans, so the money often went unspent. The government allocated no funding to help those affected back up onto their feet and, in 1927, there was no real precedent for doing so. President Coolidge was totally indifferent, refusing to involve himself in flood relief or even to visit the disaster zone despite pleas from local politicians and his own Cabinet. Yet the conditions on the lower Mississippi were so horrendous that the public and the press began to wonder whether the victims needed direct aid from the federal government.

When the floods finally receded, sharecroppers planted what the Red Cross gave them to plant only to see these new crops (Hoover's soybeans chief among them) collapse under an onslaught of poor weather and voracious insects. The disease pellagra, caused by a diet chronically deficient in niacin, had always visited the poorest Southern farmers during the winter. Now it afflicted fifty thousand people, laying them low with fatigue, shooting pains, and hallucinations.[21] A Red Cross report documented extensive fraud and racist inequity in the relief efforts. Hoover didn't want to hear about it, but, needing Black support in the Republican Party, he decided to string them along. To prominent Black leaders he floated the possibility of land reforms, using Red Cross funds to restore ownership to Black sharecroppers who had been forced to sell off their land for pennies when Reconstruction failed at the end of the nineteenth century. Hoover knew the plan was a long shot, and his Black allies did too, but they hoped that maybe when Hoover became president he'd follow through.

For the present Coolidge was still in the White House. Reluctantly he signed the Flood Control Act of 1928 to spend at least $300 million and as much as $1 billion on a massive system of comprehensive flood control for the lower Mississippi River. Sixty years after Charles Ellet's momentous report calling for the same, the federal government finally planned out a network of dams, reservoirs, outlets, and levees to protect the great funnel at the nation's heart. The federal government was officially in the business of interstate infrastructure—a role it had historically resisted but has never since relinquished.

The Mississippi Flood was the biggest news story of 1927, bigger than anything Charles Lindbergh did. Bessie Smith, the renowned singer from Tennessee, recorded "Homeless Blues" in 1927. Speaking for those displaced, Smith sang: "Mississippi River, what a fix you left me in . . . Ma and pa got drownded, Mississippi you the blame." The last line of her song hints at what many Black sharecroppers chose to do: "I'm gonna flap my wings and leave here, and never come back no more!"

Northern cities like Chicago offered jobs in heavy industry that yielded better wages than sharecropping without many of the risks, and the failure of Reconstruction in the South meant Northern states offered marginally more political rights to Black Americans. Of course migrants faced constant discrimination in the North as well, but enough felt their fortunes had improved to inspire subsequent waves of migration. The Great Flood of 1927 accelerated a movement already well underway.

The dispossessed people of the lower Mississippi settled in St. Louis, in Cairo, Illinois, in Peoria, in Detroit and Cleveland, and most often in Chicago. The U.S. Census counted 44,103 Black Chicagoans in 1910; by 1930 that number was 233,903. The Great Migration, as it has come to be known, was one of the largest internal migrations in world history, and it forever remade the United States. The Chicago jazz and literary scenes, the sounds of Motown, the playhouses of Cleveland: one cannot imagine the cultural movements of the twentieth century without the contributions of Black migrants and their descendants. When the waters rose, free people naturally sought higher ground. Herbert Hoover became president of the United States in 1928 and promptly lost interest in the delta reconstruction project that had consumed him the year before. The empire Will Percy meant to inherit from his father fell to a quiet shambles, drained of cheap labor and lacking the eager capital needed to rebuild.

Emphasis and Arrangement

The Great Depression that would descend during Hoover's presidency left most of the South totally devastated. Then, as now, Southern governments took pride in their lack of safety nets and regarded material suffering among the poor (of every race) as a sort of reflected virtue. Most of their citizens had no political rights to speak of and thus no power to alter their circumstances. When Franklin Roosevelt took office in March 1933 his inaugural address emphasized the need for urgent action to improve material conditions among the nation's people: "I assume unhesitatingly the leadership of this great army of our people dedicated to a disciplined attack upon our common problems." He also noted the broad authority given him to affect those policies, the malleability of the U.S. Constitution in the hands of a determined president: "Action in this imagine and to this end is feasible under [our] form of government. . . . Our Constitution is so simple and practical that it is possible always to meet extraordinary needs by changes in emphasis and arrangement without loss of essential form."[22]

Though his opponents attacked him for decades as a socialist radical, neither word accurately described Franklin Delano Roosevelt. Born wealthy and politically connected, Roosevelt cultivated a reputation as a traitor to his class that served him well in politics but was always overstated. He did not want to demolish capitalism. On the contrary, Roosevelt was a fervent believer in the American system of commerce, who sought to amend its worst excesses so as to thwart the rise of socialism domestically. The Great Depression had cost that system its legitimacy, and Roosevelt understood the dangers of rising popular discontent. In order for the rich to keep getting richer, the poor needed to see a broader correction. That isn't to cast Roosevelt's populism in a cynical light; the man was genuinely driven to help his countrymen. Roosevelt understood the scale of the task before him, the ruin left behind by Hoover's insistence on treating the Depression-stricken United States like the flood-stricken Mississippi Delta. Once again Hoover had trusted business elites to save the day, to provide credit, to hold up a bargain those elites did not believe existed. Roosevelt knew that conquering this catastrophe would require a Cabinet of men and women who would pursue the people's business with independence and moral clarity.

Frances Perkins was born to an educated Boston family in 1880 and followed the women of her family into the teaching profession. She became an ardent suffragist and labor activist in New York City, drawing the attention and encouragement of both former president Theodore Roosevelt and his cousin Franklin Roosevelt, then a New York state senator. When the younger Roosevelt became governor of New York in 1929, he appointed Perkins to be the state's very first industrial commissioner. When he won the presidency in 1933, he appointed Perkins to be his secretary of labor. No woman had ever held a Senate-confirmed Cabinet position before, but Perkins would hold the position until after her boss's death in 1945. In *The Roosevelt I Knew*, her memoir of serving the administration, she recalls telling Roosevelt before her confirmation that she intended to push for the most radical agenda in the nation's history: forty-hour work weeks, a minimum wage, income supports for injured workers, paid benefits for the involuntarily unemployed, a ban on child labor, and universal government pensions for the elderly. She laid all this out hoping Roosevelt would choose someone else. He confounded her by heartily endorsing her ideas. She adjusted her approach: "'But,' I asked, 'have you considered that to launch such a program we must think out, frame, and develop labor and social legislation, which might then be considered unconstitutional?' "'Well, that's a problem,' Mr. Roosevelt admitted, 'but we can work something out when the time comes.' And so I agreed to become Secretary of Labor after a conversation that lasted but an hour."[23]

Roosevelt took the plunge, and that risk won him the most important secretary of labor in the nation's history. Frances Perkins never enjoyed the public regard accorded some of her colleagues, but one cannot imagine the modern United States without the labor and social reforms she championed. Perkins bickered constantly with Roosevelt, always pushing for more than he was comfortable giving her, yet he considered her one of his dearest allies. The Roosevelt administration was a fascinating machine, a hanging mobile whose shape was sustained by tension between its elements. The most progressive of those elements did not always get their way, but neither did they break faith with Roosevelt.

Bolstered and encouraged by his Cabinet, Roosevelt immediately began pumping federal dollars into infrastructure. As a first-order proposition this made sense, as the vast nation suffering from years of economic hardship naturally abounded with projects languishing for lack of funds. Spending on roads, bridges, hospitals, dams, and levees created jobs and got money into the hands of workers who would immediately spend it on essential items and boost commerce. It would also, Roosevelt hoped, pay dividends in the long term as the finished projects could benefit their communities for decades. Many of these structures continue to serve today, nearly ninety years later. They have become features of our landscape; every part of our country would be poorer and dimmer without them. They happened because the nation was in a dire crisis its leaders were unsure it would survive, and that crisis gave the federal government license to undertake valuable projects it had previously scorned for reasons of laissez-faire ideology.

Muscle Shoals

In rebuilding the nation's economy Roosevelt would rely not just on his cabinet but on his Congressional allies, among them Senator George Norris of Nebraska. Norris represented the progressive wing of the Republican Party, roving far enough from its probusiness core that he left the party and won reelection in 1936 as an independent caucusing with the Democrats. No man living fought harder for public utilities and the public ownership of resources. Because of his influence, every electricity provider in the state of Nebraska has been publicly owned since 1940. As a U.S. senator, Norris pushed hard to retain federal holdings of land and resources in the face of a Republican Party that would have been happy to sell them off.

The Tennessee River starts at the city of Knoxville, where the French Broad River meets the Holston River under the serene pickets of the Blue Ridge Mountains. Gravity pulls the new river southwest to Chattanooga, west through northern Alabama and then due north to join the Ohio River just a few miles before it dumps into the Mississippi at the roaring gate of Cairo, Illinois. With dozens of tributaries, the Tennessee itself becomes the Ohio's largest tributary. It drains tens of thousands of square miles across seven states, almost all of which receive heavy seasonal rainfall, and, for that reason, the Tennessee River loosed regular floods nearly as deadly as the Mississippi. The victims in the Tennessee Valley resembled those in the Mississippi Delta: poor sharecroppers, most of them Black, living in tar-roofed shacks without electricity or indoor plumbing. A tempestuous river at the best of times, the Tennessee's wide waters concealed dangerous rapids and shallows to impede navigation. One of these places the steamship pilots learned to fear more than the others: a cauldron of hydraulic violence in the Alabama stretch known as Muscle Shoals.

That hazard made the spot appealing as a dam site. Raise the water level, slow the flow and the shoals would disappear entirely. Meanwhile, the vertical drop at the site yielded the most hydroelectric potential of any site east of the Rockies, but the project would impact multiple states, and thus only the federal government could act. World War I inspired Woodrow Wilson's administration to invest in domestic weapons production, and, in 1916, they picked the Muscle Shoals for a major facility. Munitions factories demanded nitrates to make explosives, so the government planned a pair of nitrate plants. To power them they would build a large hydroelectric dam spanning the entire 4,500-foot width of the river, vaulted high above the banks on arches meant to evoke the ancient aqueducts of Rome.

The Army Corps of Engineers began work in 1918, employing over 18,000 workers. The end of World War I drained some urgency and government funding, delaying the dam's completion until 1924.[24] With only 40 percent of its turbines installed and the nitrate plants in need of conversion to civilian industry, the newly christened Woodrow Wilson Dam was rich with potential but was something of a fixer-upper project.

Henry Ford wanted to buy up the whole place, bring in heavy industry, convert it into a Ford company town: the Detroit of the South. President Coolidge likely would have let him have it, and President Hoover as well but for the ferocious resistance of

Senator Norris. Any time a bill to sell the Wilson Dam bubbled up through Congress, Norris took it down with the surety of a shrike. He spurned Ford's offer of $5 million (the dam had cost $130 million to build) and instead proposed a solution based on his belief in comprehensive government-based resource management. The Norris Plan proposed finishing the Wilson Dam's turbines, converting the nitrate factories to produce cheap fertilizer and selling the dam's huge hydroelectric surplus to homes and businesses throughout the region. The revenue from Muscle Shoals could then be used to fund the construction of more hydroelectric dams on the Tennessee River and its most important tributaries. It was an impressively airtight approach to domestic resource development, and Norris successfully shepherded his plan through both houses of Congress only to have it vetoed by both Coolidge and Hoover. Frustrated, Ford abandoned his Alabaman ambitions. But neither could Norris see his own goals realized.

Roosevelt's election resolved the impasse. He saw in the Norris Plan a ready-made scheme scaled to the size of the nation's problems and pounced on the opportunity. The structure his administration worked out with Norris would serve as a model for public resource development across the United States and the world: the Tennessee Valley Authority (TVA). Premised on Norris's belief in integrated resource development, the TVA was, in Roosevelt's words, "a corporation clothed with the power of government but possessed of the flexibility and initiative of a private enterprise." Congress passed Norris's bill on May 18, 1933, just two months after Roosevelt's inauguration. Norris described the day as a dream come true.

The TVA had plans for the whole river valley, but Muscle Shoals was the key to making it all work. "Muscle Shoals gives us the opportunity to accomplish a great purpose for the people of many States," Roosevelt said in Montgomery, Alabama during the interstice of his election and inauguration, "and indeed, for the whole Union. Because there we have an opportunity of setting an example in planning, not just for ourselves but for the generations to come, tying in industry and agriculture and forestry and flood prevention . . . into a unified whole over a distance of a thousand miles." The president and his advisors expected the TVA to do good work in the short term, but they also saw it as one component of the powerful and responsive central government they meant to build. The TVA would make it easier to expand state capacity in the future, by demonstrating to a suffering public that government could meaningfully improve their lives.

Light and Heat

It would do this by electrifying the Tennessee Valley. In 1933 barely 10 percent of American farming households had access to electricity.[25] No technological barriers held back electrification, as anyone in proximity to a city had enjoyed electricity for decades. But power companies did not want to invest in new plants in rural areas without guaranteed customers, and farmers did not want to invest in electrifying their homes and businesses without assurance that prices would stay low. It was a straightforward case of market failure. The TVA sought to resolve the impasse by creating new impasses on the Tennessee River and its tributaries: a network of hydroelectric dams to inundate the region with cheap power while safeguarding it against floods of the literal sort.

To administer electric policy for the new organization, Roosevelt appointed David Lilienthal, a lawyer who spent his early career antagonizing private utility companies in Illinois and Wisconsin. A civil engineer with a passion for community organizing named Arthur E. Morgan served as chairman and oversaw dam policy. Agricultural policy went to the confusingly named Harcourt Morgan (no relation to Arthur). These passionate and stubborn men fought like dogs—especially Lilienthal and Arthur Morgan, the pragmatist and the idealist respectively. Harcourt Morgan generally sided with Lilienthal, leading to Arthur Morgan's departure in 1938, but the foundation the three laid accomplished an undeniably historic transformation in the Tennessee Valley. A decade after Roosevelt established the TVA, Arthur Morgan had built sixteen new dams and renovated five others. Long stretches of the Tennessee River became navigable for the first time in history once the dams raised water levels and TVA crews had dredged clean channels. Farmers eagerly moved into hundreds of thousands of acres of prime black-soil bottomlands guarded against floods for the very first time. Harcourt Morgan established "demonstration farms" to teach sustainable farming practices. Simultaneously he oversaw the Muscle Shoals nitrate plants, now producing 150,000 tons of fertilizer each year.

One of the first new dams the TVA built bottles up the Clinch River, which rises in the hills of Virginia before sliding southwest down the Great Appalachian Valley into the state of Tennessee. It collects considerable tributary water and speed on its three-hundred-mile journey to join the Tennessee River west of Knoxville. Historically the Clinch hosted some of the richest shoals of river mussels in the South (the name "Muscle Shoals" may in fact refer to local shellfish, not the rapids). For thousands of years Indigenous Americans of the Cherokee Nation congregated along the banks to steam and devour mussels. They adorned their clothes and hair with river pearls, they built their homes with river cane, they encouraged chestnut and mulberry trees to grow tall by burning or pruning out competitors, such that Hernando de Soto's 1540 expedition encountered forests of dreamlike bounty. Every Indigenous village presented the Spaniards with so much produce that even their horses grew fat. De Soto gave the locals no credit for millennia of horticulture, marveling the mulberries "grow wild in the fields without being planted or manured and are as large and vigorous as if they were cultivated in gardens."[26] In the fertile bottomlands the Cherokees grew beans, corn, and squash in fields so vast they stretched unbroken from one riverbank village to the next.

The Clinch takes a steep fall just a few miles from its confluence with the Tennessee River—a spot engineers noted was ideal for a hydroelectric dam. The May 1933 bill establishing the TVA also authorized a hydroelectric dam on the Clinch River, work on which began immediately. They hired Hungarian-born architect Roland Wank to oversee the construction according to the design already completed by the Corps of Engineers and Bureau of Reclamation. Wank tweaked the blueprints to comport with his preferred modernist style, and, less than a year later, had finished construction on a magnificent dam—1,860 feet long, all slopes and angles, tucked sleekly into a gap between two round hills. Wank's vision, controversial at the time as it contrasted with neo-Classical designs like the Wilson Dam, came to be seen as state-of-the-art. Wank was personally rewarded with a lengthy tenure as the TVA's head architect.

They named the dam for Norris, who guarded the Tennessee River like his own kin and who vanquished Henry Ford and made the TVA possible. The Norris Dam was so grand and visually daring it attracted domestic tourists and foreign luminaries alike, all drawn by this work of geoengineering as art: Jean-Paul Sartre, Jesse Owens, and Jawaharlal Nehru all visited Norris in its early years. Yet for all the acclamation the great dam erased a great deal of living history. For generations no one has known the river in its natural state. The mussel beds, which fed the Cherokees and once made the Clinch River one of the United States' biggest producers of river pearls, are gone along with the river cane. Today there are twenty-nine dams on the Tennessee River and its tributaries, each of them impounding the natural flow of water and sediment demanded by shellfish.

David Lilienthal put the power from those dams to work—a million kilowatts of generating capacity by the start of World War II. The private sector indignantly opposed the TVA from the outset, denouncing it as a socialist enterprise. Wendell Willkie, president of the competing Commonwealth & Southern (C&S) utility conglomerate, insisted the TVA would never find sufficient electric demand to justify its budget. As it turned out, once promised all the electricity they could use at some of the lowest rates in the United States, farmers finally made the long-avoided switch to electric. For the very first time many people in the Tennessee Valley could light and heat their homes at night, the electricity provided by the very same dams that kept their fields from flooding every spring.

The free market was content for them to live without this basic modern necessity indefinitely, as Willkie's objection proved—they didn't even believe there was demand for electricity. If the federal government had not assumed responsibility for electrification, it simply never would have happened. In the end, even Willkie's company benefitted from the TVA as the last-mile infrastructure they built (poles, wires, local transformers, and house-by-house hookups) carried C&S voltage as well. Even the New Deal's foes made out like bandits in the electric bonanza, described in historian Michael Hiltzik's very impressive book, *The New Deal: A Modern History*: "Across the Southeast, rates fell so sharply that residents and businesses started thinking up new ways to use electricity—a situation that had been unimaginable only a few years earlier. . . . New electric appliances began appearing in homes—refrigerators, washing machines, water heaters—due in part to a TVA program subsidizing their purchase. . . . by 1936, the demand for electricity from Willkie's company was so great that [C&S] could not meet it without purchasing excess generation from the TVA."[27] Electric appliance ownership tripled even as power rates fell 30 percent lower than the national average. Willkie, in the parlance of our times, did not know how to "take the W." He declared for president in 1940 on a platform opposing the New Deal, won the Republican nomination and proceeded to carry just ten states while losing to Roosevelt by ten percentage points and five million ballots nationally.

Following the Pearl Harbor attack in December 1941, Roosevelt's administration eagerly enlisted the TVA to bolster the war effort. The Tennessee River was not nearly so large as the West's mighty Columbia River, but the TVA ran so many dams and turbines that the New Deal's experimental corporation was now the number one energy generator in the United States.[28] The fight against Nazi Germany would demand more. Freed from any budgetary constriction, the TVA built on the Tennessee with a frenzy that surpassed even the go-go years of Roosevelt's first term: twelve new hydroelectric dams in addition to an arsenal of industrial projects powered by the extant dams. Aluminum smelters appeared alongside aircraft factories, chemical plants, and shipping facilities. The TVA employed twenty-eight thousand people in engineering and construction alone; the factories represented tens of thousands more. A total of 60 percent of the phosphorus used to fire American munitions downrange originated in TVA plants that melted down phosphate-rich ore around the clock. A total of thirty-five hundred lumber mills cranked out components for every wooden product the Army used, from furniture to gunstocks. The Tennessee River, once scarcely navigable, became a 650-mile highway for rafts of barges hauling materials to and from the plants. TVA engineers even accompanied the army overseas to assist with surveying and mapmaking.[29] Turned to wartime ends, the river corridor became a leviathan of industry. Drawing endless strength from the Columbia in the west and the Tennessee in the east, the United States rose among World War II's belligerents as the world's premier industrial power.

Bloody Arithmetic

To men like Herbert Hoover and Wendell Willkie, the notion that money spent on infrastructure and internal improvements translated directly into enduring economic benefits—that in certain circumstances the government could simply step in and succeed where the private sector had failed—seemed a trick, a scam. Roosevelt appeared to his opponents like Wile E. Coyote, sprinting off a cliff and out into empty air. Surely if the contradictions were illuminated, he would be forced to look down, and the illusion would be dispelled. He would simply plummet. That this failed to happen, time and again, enraged Roosevelt's opponents even more than his "socialist" agenda.

At the same time he fell short of the left's highest expectations. Democratic politicians in the South were some of the nation's most vicious white supremacists, and they represented a substantial cohort of Roosevelt's Congressional majority. They had no interest in seeing economic or social conditions improve for Black Americans, especially those on whom the ownership class relied for cheap labor. As a result many New Deal initiatives were deliberately watered down, designed to exclude nonwhite people where possible: most notably the first incarnation of social security exempted agricultural and domestic workers from old-age and survivors' benefits. The Civilian Conservation Corps, responsible for so many construction projects nationwide that at least one is probably still standing in your hometown, began as a racially integrated program but was later segregated in the South as a response to persistent local outrage. Roosevelt's Federal Housing Administration advanced segregation by pushing Black citizens into urban centers while excluding them from the suburbs. The GI Bill paid for millions of World War II veterans to go to college, but Rep. John Rankin of Mississippi (a cosponsor of the original TVA bill) insisted that local organizations distribute the money and thus functionally locked Black veterans from the South out of their benefits.

We must also acknowledge the single worst act of Roosevelt's presidency: Executive Order 9066, signed in February 1942 to authorize the expulsion of over one hundred thousand people of Japanese ancestry from their homes. Though completely innocent of sabotage or espionage (not a single Japanese American or Japanese resident of the United States was ever convicted of these crimes), these families were forced to wait out the remainder of World War II in internment camps scattered across the nation's most inhospitable places. Many lost their homes, businesses, and property to opportunistic vultures and had to begin again from nothing when they returned. They would receive no compensation until the 1990s, after most of those wronged were deceased.

For all of these sins Roosevelt surely deserves history's criticism. Yet it would be myopic to claim, as some do, that the New Deal represented a retrograde step in the nation's awful racial history. Certainly Black Americans deserved more from the New Deal, but the material benefits they claimed were vast compared with those available before Roosevelt's election. What's more, their passage and popularity made possible almost every genuine advance in American governance that has occurred since the 1930s. Prior to Roosevelt, the U.S. government *refused* to spend on universal programs, on internal improvements. Over and again American presidents *refused* to spend on infrastructure even if it was badly needed, even if it was popular. Because of pure ideology, they refused. In his very first term, Roosevelt swept all that away. Crises may push even pragmatists to extremes. As a result of Roosevelt's extremism, entire swaths of the country were transformed—often the poorest and hardest-suffering communities in America, communities no utility company would touch with a thirty-foot pole. Tennessee freeholders who had never had electricity could turn their lights on. Mississippi Delta sharecroppers could raise their crops without Biblical destruction erasing everything they'd ever worked for. In the spring of 2021, the Tennessee River is in the process of weathering the rainiest winter on record—the kind of precipitation that would have inundated half the valley a hundred years ago. Instead the dams hold. Floods may still occur during extreme weather events, which are growing more frequent due to climate change, but the regular seasonal flooding that defined the region for centuries has ceased. Millions of people get to live in a comfortable modern society because in 1933 a team of decent people took a risk on doing the right thing. The New Deal's lasting impact on the eastern United States can be summed in one sentence: the South no longer floods.

4

A TOWERING HEIGHT

Harold LeClair Ickes was born in a completely different world from the one he would one day attempt to lead. The mighty Pennsylvania Railroad founded the city of Altoona as a company town in 1849, betting that its position in the hills of central Pennsylvania would pay dividends. Indeed the Civil War spiked demand for locomotives and the Pennsylvania Railroad was more than happy to expand their capacity to win government contracts. Two thousand workers lived in Altoona in 1854, but by 1865 that number had quadrupled, and by 1890 the city's population stood higher than thirty thousand. Harold Ickes was born in 1874 to a merchant who served those workers: a man named Jesse Ickes (pronounced ICK-us), a pitiable philanderer and alcoholic. Harold was the oldest of six children, so he picked up the slack. As many children in his situation might, he developed quite a dim view of his father and a diametrically opposed view of his mother—a pious woman Harold elevated in his mind to sainthood. Very early in his life, the boy everyone called "Clair" developed a moral compass that would always pull powerfully at him.

Ickes's biographer T. H. Watkins writes of Altoona's grim industrial squalor: "Many of its streets still unpaved . . . gutters running with raw sewerage, business-district buildings of bare brick, wood, or industrial stone, mundane workers' cottages bravely annotated with Victorian scrollwork, naked telephone and electric-light poles canted over the avenues . . . and over it all the stink and noise and soot of progress unrestrained."[1] Accidents in the factories and collisions on the rail lines killed workers almost every day. In 1890 Harold's long-suffering mother died due to a respiratory infection of the sort that antibiotics would one day render essentially harmless. Knowing her husband was ill-equipped to handle all six children, Mattie Ickes sent sixteen-year-old Harold and nine-year-old Amelia to live with her sister in Chicago.

In Chicago Harold would find himself, slowly and with much painful labor. He worked full-time, unpaid, in his uncle's shop whenever he was not at school, but the talented young man won the admiration of several teachers, and their encouragement was sufficient to keep him in school. Much of Chicago burned to the ground in 1871, but twenty years later it had been completely rebuilt. One enduring institution was the University of Chicago, then run by the estimable Bible scholar William Rainey Harper. Harper possessed a strong Progressive bent despite his tight connection with John D. Rockefeller (the university's principal funder) and ensured working-class young people of talent could gain admittance. Crucially, he also set them up with work to pay their tuition and living expenses. To Harold Ickes, the University of Chicago offered his only real opportunity to start a life for himself.

He did not excel but managed to pass his courses. A bog-standard bout of unrequited love put Harold in an adolescent melancholy for which he sought the refuge of sore losers the world over: electoral politics. He became president of the University Republican Club and fell in with the mayoral campaign of an anticorruption candidate. Ickes's man lost, but the moral drama and high stakes of Chicago municipal politics lit a fire in the youngster that would burn the rest of his days. As Watkins writes, "absolute Good locked in sweaty combat with absolute evil . . . to [Ickes's] whole-souled and unsophisticated intelligence, it all smacked of the very guts of life."[2]

Needing a job after graduation and qualified for the first time in his life for something other than manual labor, Ickes tried his hand at newspaper reporting. Newspapers were big business in an era where every big city boasted a half-dozen thriving dailies, so Ickes got a junior reporter job and moved from one rag to the next, gradually ratcheting up his pay. He found the hidden life of Chicago intoxicating; he was fascinated by the brute force of political machines, on daily display. "It was the plug-ugly, the heavy-handed manipulator of ballots, the man who could smell out the loot and get to it first, who was in the saddle," Ickes later wrote in his memoir, *Autobiography of a Curmudgeon*.[3] "I discovered, from my place in the press box, that the ebb and flow of surface political sentiment did not . . . give any true indication of the violent cross currents that run just a little deeper. I became cynically wise to the selfishness and meanness of men . . . I found out, too, that there are men and women who often serve the common good at great self-sacrifice and without any hope of or desire for political reward."

Covering the 1900 presidential election granted Ickes the opportunity to become acquainted with high Republican officials, including national party chairman Mark Hanna and eventual (following William McKinley's assassination in 1901) President Teddy Roosevelt. The newspaper for which Ickes worked folded shortly thereafter, and the young man took a series of jobs on local Republican campaigns. His candidates didn't win, but by the time Ickes decided to go to the University of Chicago Law School in 1904, he had earned his reputation as an honest and savvy campaign manager.

Ickes practiced law intermittently after graduating, working on cases he found personally inspiring and billing pittances to his clients. He had no reverence for the law as such. Rather, he saw the law as a tool like any other, to be used on behalf of the powerful or the powerless. At the request of Jane Addams, founder of the Hull House organization that provided for the needs of Chicago's immigrant community, Ickes represented the family of a young Russian-Jewish man named Lazarus

Averbuch. Averbuch had allegedly tried to assassinate Chicago's chief of police on his own doorstep and had been killed in the attempt, so there would be no trial, but Addams and the city's Progressive community knew the city's Democratic government would use the official inquest as an opportunity to scapegoat the Russian-Jewish community as a hive of subversives. Ickes raised $2,000 and forced the city to exhume Averbuch's body for a second autopsy conducted by independent physicians. That autopsy revealed a second crime: Averbuch's skull had been opened and his brain removed, presumably by coroners interested in gruesome pseudoscience. Ickes used this knowledge as leverage over the city government, which backed off the official elements of their crackdown.

Ickes's living expenses during that first decade of the twentieth century were largely covered by the household of his friend James Thompson, who had married into wealth. His wife, Anna Wilmarth Thompson, was the mistress of her home and ran roughshod over her doormat of a husband. She despised what she saw as his personal weakness and admired the steely determination of his friend, Harold Ickes. It was at her insistence (relayed dutifully through James) that Harold moved from his law school fraternity house into the Thompson's home in an upper-middle-class Chicago enclave. If this seems to you like a bad idea, you have better instincts than the nation's greatest secretary of the interior—a man whose tightly buttoned-up professional reputation concealed an absolute shitshow of a personal life.

Anna, like Harold, was the child of an alcoholic father. Both of them craved the security of stable relationships and would go to nearly any length to avoid severing a bond, yet neither could ever fully trust another human being. They also shared a morality that could fairly be described as Victorian: rigid, traditional, fanatically devoted to keeping up public appearances. Never could a crack appear in the façade. So it was surely awkward when the two began falling for each other. The attraction developed at least a year before either could bring themselves to acknowledge it. James was a nerd's nerd, a scholar of European history whose head was ever in the clouds, but even he couldn't help noticing. When he complained to Anna, she reminded him very tactfully that the family house was her house, the family money her money. He backed down, no doubt validating the contempt she felt for him. James's own brother wrote to Harold explaining that he was a far more impressive man than James and should feel no compunction whatsoever about partnering with Anna. James's son even preferred his "Uncle Clair" (who always held a sweet spot for children) to his aloof and distracted father.

Even once Anna and Harold acknowledged to each other that they were in love, their mutually held morality kept them from acting on it. The result was a household in the grips of agony: a collapsed marriage and an unrequited affair under one roof. They could not consummate the affair, they agreed, until Anna had found an honorable avenue for divorce. Even once that divorce was secured—one of the couple's several adopted children told Anna that James had sexually assaulted her— Anna insisted they could not marry until James had found a new wife. In 1911 they finally married and Anna quickly became pregnant, which triggered explosive fights as she felt the proximity of events implied they had wed merely to have sex with social sanction. It fell upon Harold to point out that this was the reason most people marry. He could also be incredibly stubborn and irascible; his Victorian sensibilities also led him to withhold emotionally and physically from Anna in response to any disturbance, which only escalated conflicts and encouraged them to fester.

Harold Ickes spent the next two decades working quietly on Republican political campaigns, arranging the flows of semilegal money that any politician needed to have a serious shot in Chicago. But the Republican Party was drifting rightward. Ickes considered President Warren Harding corrupt; the poor governance of Presidents Coolidge and Hoover alienated what progressives remained in the party.

Franklin Roosevelt's rise in 1932 wrapped up Ickes's loyalty to the Democrats. In his autobiography he wrote, "I had been head over heels for the nomination of Franklin D. Roosevelt for President. I had followed his career and thought highly of him, although I had never met him personally."[4] Ickes went to work for the Illinois branch of Roosevelt's campaign, which sparked an argument with his wife. Anna had branched out into her own pursuits, which included academic and charity work with Indigenous Americans in New Mexico and running for local office as a Republican. She accused Harold of tainting her good name in the party; he parried by leveraging her other interest. If Roosevelt won, he told her, he would lobby for a position as Secretary of Indian Affairs. According to T. H. Watkins, this is the first time in his life Ickes ever voiced an interest in holding federal office.

Whatever the genesis of his idea, it mollified Anna. Ickes worked his connections in the Roosevelt campaign to put his name forward for Indian Affairs, but competitors had inside tracks for the job, and Ickes shifted his sights to the position of first assistant secretary of the interior. Before long he had convinced himself he was qualified to run all of Interior. Even if he had never been a bureaucrat, he had two decades of wrangling under his belt in one of the nation's most ferocious political environments. Even if he had no experience with land or natural resources, he had supported Teddy Roosevelt's Progressive agenda and thus understood the basic principles of conservation. These are the sorts of things one tells oneself to make it through the application and interview process for a job one does not really deserve.

And yet it worked. Ickes was lucky; F. D. Roosevelt wanted someone from Teddy's Progressive faction to run Interior, and his first two picks both declined the job. Watkins writes, "Six months earlier, Ickes had not even known he wanted this job;

now he wanted it possibly more than anything he had ever wanted in his life."[5] The desperate Harold wired everyone he knew in Democratic and Progressive circles, begging and wheedling for a good word with Roosevelt in a way he never would before or after. For weeks Roosevelt sought a high profile candidate, but none emerged, and, in the end, he told Ickes in person at the Roosevelt Hotel in New York City that he would be the nation's next secretary of the interior.

The Great Campaign

Roosevelt meant to begin his work immediately. Just hours after the new president's inauguration, Vice President John Garner left the festivities and called the U.S. Senate to session. With scarcely any deliberation the body confirmed all ten of Roosevelt's cabinet nominees, and, that evening, the newly empowered bureaucrats gathered at the White House for a mass swearing-in, the first of its kind in the nation's history. Harold Ickes was suddenly one of the most powerful men in the United States.

Anxiety, not exuberance, characterized the national mood on Inauguration Day. Roosevelt had nearly been assassinated seventeen days earlier in Chicago by an Italian immigrant named Giuseppe Zangara who had to stand on a chair during the shooting on account of being only five feet tall. Zangara missed Roosevelt and instead killed Chicago mayor Anton Cermak. More pressing to the average American was Inauguration Day dawned amidst a banking crisis in full bloom: more than three years after the first crashes of 1929, nearly every major bank in every state had already faced runs on their cash deposits, and now terrified investors were cleaning out their gold reserves as well. The federal government did not insure bank deposits in 1933, so the Treasury Department was left scrambling to find enough liquidity to shore up the nation's banking system against total collapse. Business leaders like Henry Ford saw no reason to incur risk on anyone else's behalf. The outgoing President Hoover had spent his lame-duck months sulking and trying to convince Roosevelt to renounce the policies that had just won him a landslide election. Treasury begged Hoover to declare the banks temporarily closed so that they might avoid insolvency, but Hoover refused. Told he was fiddling while Rome burned, the petulant and defeated Hoover snapped back, "I have been fiddled at enough and I can do some fiddling myself."[6]

Once in office Roosevelt acted swiftly, ordering banks around the country to close indefinitely. He drew on presidential powers enacted during World War I: the Trading with the Enemy Act of 1917 had been meant to limit banks loaning money to enemy nations but gave the president nearly dictatorial power over the financial industry if he chose to exercise it. Even powerful Democrats like Senator Carter Glass of Virginia insisted Congress had not intended such an outcome.[7] Roosevelt didn't care. On paper the authority was his, and any legal challenge would unfold too slowly to stop him. The banks closed, and, for more than a week, the people of the United States simply went without money. Those who had cash on hand were free to spend it, but laborers were paid largely in company scrip.

The emergency measure kept thousands of banks from going under, but they still needed guarantees of liquidity before they could open again. Roosevelt instructed Congress to pass his Emergency Banking Act (EBA), empowering Treasury to backstop banks around the country. Since the Treasury could simply issue notes at will, the EBA meant that every deposit in every bank in America was suddenly insured by the government. That was an unbeatable deal almost anywhere in the world, and investors leapt to take advantage. By the end of March 1933 Americans had deposited $1.2 billion back into the banking system.[8] The Treasury barely had to touch its slush fund as the financial system righted itself with such speed that Roosevelt's people dubbed their maneuver a miracle.

Early success emboldened Roosevelt and his cabinet. They

faced rolling catastrophes in almost every sector of the U.S. economy, from manufacturing to agriculture. Twelve years of rule by right-wing Republicans left the country in a disastrous state. Speaking specifically to the issue of collapsed farm incomes due to collapsed prices, Roosevelt articulated a principal he put into practice throughout his administration: results came first, methods second. "If we cannot do this one way we will do it another. Do it, we will."[9]

Harold Ickes cultivated a public reputation for impeccable honesty, but when it came to wielding power he was as cunning and driven as anyone in Washington. He installed friends and family members throughout the Department of the Interior, picking roles where they would not embarrass him, and in so doing he established a trusted network of internal informants. His underlings even surveilled other Cabinet-level officials, especially Ickes's personal rival, Secretary of Commerce Harry Hopkins.

Both men shared a ferocious commitment to improving the lives of working- and middle-class Americans, but their differing temperaments—Ickes a boiling stove of consternation and insecurity, Hopkins sardonic and condescending—caused constant friction. They argued over methods, too: Ickes focused on large enduring programs and battling corruption in government whereas Hopkins funneled public money to workers as fast as possible. Most galling to Ickes was Hopkins's close personal relationship with Roosevelt, with Hopkins even moving into the White House in 1940. Ickes dealt with Roosevelt the same way he did everyone else: by roaring and whimpering in turn, showing his pugnacious side before an inevitable cycle of withdrawal and reconciliation. When Ickes felt slighted by Roosevelt, he would threaten to resign until the president performed the emotional labor of soothing him, smoothing his feathers, and assuring Ickes he was a valued member of the cabinet. Hopkins took setbacks in stride and used his connections with Roosevelt's inner circle to bring his boss around.

With the benefit of hindsight, even Ickes conceded Hopkins won their arguments more often than not: "While in a contest with Harry I might win the President's mind, his heart would remain with Harry."[10] Roosevelt, for his part, saw the two men as perfectly complementary lieutenants and struggled in vain to help them understand one another.

Hopkins's legacy can still be seen around the country, in thousands of public buildings erected by his Civil Works Administration and later the Works Progress Administration (WPA) (not to be confused with the Public Works Administration [PWA], which worked under Ickes's authority). One cannot reasonably doubt that Hopkins's projects transferred huge sums of money to the nation's working class. But Ickes, tasked with the very earth of the struggling United States, held ambitions more grandiose than did Hopkins. He had been gifted an extraordinary opportunity and worked always with an eye to posterity. Under his direction the PWA repaired or replaced more than five hundred school buildings in southern California following a March 1933 earthquake. "Most of them," historian Michael Hiltzik notes, "built to the most exacting seismic standards of the time, are still in use."[11] Bridges went up across the country, bringing isolated communities like Florida's Key West into the national fold for the first time. Roads, sewage pipes, power lines, train stations, airports—the government built all of these with unprecedented abandon, and most of those facilities remained in service for decades if not until the present.

Roosevelt offered Harold Ickes the funds to design and build a new building for Interior, and Ickes jumped at the chance. He took such pride in the architecture, the fine details, that when the time came to sit for his official portrait (painted by Henry Salem Hubbell, a family friend) Ickes arranged for the building's plans to sit on his desk.

The final product was twenty-two hundred rooms of administrative splendor, all insulated from the region's marshy summer heat by central air conditioning—the first federal office building to deploy the technology. Ickes vetted every piece of artwork displayed in the building, insisting at one point that a generic earth-water-air-fire piece was inappropriate since the Department worked to prevent floods and was "the one Department in the Government that owns no airships."[12] He asked that Indigenous and Black Americans be emphasized in the artwork, portrayed as dignified human beings afflicted unjustly by oppression.

During this time he became functionally estranged from Anna—she spent most of her time in New Mexico—and this ironically granted their relationship a degree of tranquility. Ickes started up an affair with a younger woman in his office. He felt guilty about it the whole time and finally experienced a degree of heartbreak when his paramour eloped to marry her fiancé. His personal life took a darker turn in 1935 when Anna died in a car accident; his stepson, Wilmarth (Anna's child with James Thompson), died by suicide a year to the day after his mother's death. Ickes, always messy, started dating the younger sister of Wilmarth's widow. Though thirty-nine years separated the pair, they made a good match and got married in 1938 before raising a pair of children. In his golden years Harold Ickes seems to have finally found the stable, happy family he always craved.

His professional life stayed combative. It takes a certain type of man to name his own memoir, *Autobiography of a Curmudgeon* after all. Ickes envisioned the office he built as the beating heart of an Interior Department that would one day become something even greater: a Department of Conservation empowered to drive national objectives across every acre of federal land. It was an intoxicating prospect for Ickes, having entered politics as a Teddy Roosevelt progressive. He began his campaign with the national parks, a system of hallowed natural grounds Teddy had established and which stood on rock-solid political foundations. The Secretary of the Interior regarded nature conservationists as a passionate faction so splintered by local projects that they rendered themselves powerless. Ickes had grown up in rural Pennsylvania at the cusp of the Industrial Revolution and made many trips to the West's natural splendors; he spoke movingly on a 1934 radio broadcast of his awe at visiting Glacier National Park in Montana. His love for the natural world was bone deep, yet one must note that many of Ickes's dam-and-reservoir projects also paved over large swaths of natural splendor. T. H. Watkins writes, "It was a contradiction that apparently raised no questions in his own mind," though of course conservationists often found it infuriating.[13] Harold Ickes could be your best friend or worst enemy, depending on the day and the issue at hand. He contained volumes.

From Mount Olympus to the Valley of Kings

At a time when Franklin Roosevelt's administration was looking for any excuse to build new roads and highways, Ickes fiercely opposed laying new roads through the national parks. Businesses like Yosemite National Park's Ahwahnee Hotel he considered mistakes on the department's behalf; amenities, he believed, should be kept minimal to encourage exploration on foot and by horseback. He believed ardently that the U.S. Forest Service, then as now a wing of the Department of Agriculture, should be part of the Department of the Interior and fought unsuccessfully for years to make it so. Upon arriving in office President Roosevelt reorganized the Executive Branch and, among his many changes, was shifting management of national monuments from the Forest Service to the National Park Service. Ickes recognized in the National Park Service a powerful tool—a fulcrum to transfer acreage and thus power from Agriculture to Interior, if only Ickes could win the internal battles.

He made his first offensive in the Pacific Northwest, on the Olympic Peninsula. The peninsula is a mountainous protuberance jutting into the Pacific Ocean. If you look west from Seattle, across Elliott Bay, you'll see the Olympic Mountains laid out in a wide band, a jagged snow-capped diadem whose crown jewel is the setting sun. They are old volcanoes, long dormant and badly eroded by constant precipitation. Storms from the northern Pacific whip against their sea-facing slopes, making the Olympic Peninsula one of the wettest places in the continental United States. It is a temperate rainforest: mosses, ferns, lichens, and conifers gnarled into impassable jungles of primordial green. The peninsula feeds and harbors a large population of Olympic Elk, the largest species of elk in North America, known for its striking black-and-tawny markings and great stature. Bulls can grow to over one thousand pounds.

Teddy Roosevelt established the Mount Olympus National Monument in 1909 specifically to safeguard these priceless animals. Unregulated hunting and logging had cut deep into elk territory, and their numbers were flagging. Nobody could expect the man who led the Bull Moose Party to stand by while magnificent ungulates were threatened, and Teddy converted a large parcel of federal land into protected territory (abrogating treaties with the Quinault Nation, whose reservation abuts the park). At that time the Forest Service generally deferred to local business interests on questions of conservation, and the Pacific Northwest was lousy with official corruption. In 1915 the Forest Service used the looming prospect of war to convince President Woodrow Wilson that Mt. Olympus held huge manganese deposits, and 292,000 acres of protected land needed to be opened for mining.[14] There were no ore deposits, as surveyors would discover, but the fiction allowed the Forest Service to open up that acreage for logging.

In 1933 the Forest Service sanctioned a four-day hunting season on national forest lands that caused the slaughter of roughly two hundred large bull elk—an atrocity that elicited a powerful reaction among conservationists. Rosalie Edge, a wealthy socialite who made her name first as a suffragette and then as a radical conservationist, deployed her Emergency Conservation Committee (ECC) to push for a full national park in the Olympic Mountains. She and her allies won the attention of U.S. National Park Service director Arno Cammerer, who appointed a committee to study the subject. Cammerer reported to Harold Ickes, and the labor secretary always kept a finger in every pie. The secretary personally pushed to include the Quinault River Valley in the park, and Cammerer's final recommendation sought to annex 400,000 acres of national forest land in addition to the existing 328,000 acres of the National Monument.

The Forest Service went berserk, refusing to cede any of their portfolio to the Park Service. Ickes had at least one shouting match with Secretary of Agriculture Henry Wallace over the matter, and it became increasingly clear to the irascible Interior secretary that this was a contest of power, not persuasion. Journalist Irving Brant, Rosalie Edge's ally on the ECC, lunched with Forest Service chief Ferdinand Silcox. At their meeting Silcox agreed with every point Brant raised yet stood unmoved. As Brant wrote to Ickes, Silcox stated "he would not sanction the transfer of a single acre to the 'administrative methods' of the Park Service."[15]

It was not until 1937, after Roosevelt had won reelection in a forty-six-state landslide, that Ickes pulled out his big irons. The president made a personal visit to the Pacific Northwest and the Olympic Mountains his cousin Teddy loved so well, taking the opportunity to officially rename the Olympic elk as the Roosevelt elk. It is still called that today. Franklin Roosevelt had a brilliant understanding of sentiment and symbols—what today we call marketing. Tying his family's name and sterling reputation to this huge, beautiful, threatened animal was a masterstroke. On June 16 of that year, on the last minute of the last day of the legislative session, Congress authorized the establishment of Olympic National Park. Ickes was so impressed with Brant's work that he offered the journalist the directorship of the National Park Service, confiding in him that "Cammerer has no guts. He is afraid to disagree with me."[16] Understandable, given that Ickes abused those who disagreed with him. Brant politely declined the job, but the two would soon find themselves allied again.

More than ten thousand years ago the very first Americans explored the West Coast. These were tough, resourceful people who could survive anywhere they had access to fresh water. Hiking overland, paddling kayaks or canoes, they pushed inland from river mouths to find new territory. In California they found the land shaped like a basin, the Central Valley a glacier-carved natural channel for water. Yearly melts from the Sierra Nevada ran into the valley and ended up in one of two "pipeline" rivers: the Sacramento, starting in the northern Sierra Nevada and running south; and the San Joaquin, starting in the southern Sierras and running north. Moving south, following the San Joaquin upriver, those first explorers would have eventually happened upon a region of marshlands surrounding a vast lake: beaver, elk, and dense tule reeds to the limit of their sight.

Tulare Lake, on the sunblasted plain south of modern Fresno, existed on a serendipitous patch of topography. A shallow ridge just south of the San Joaquin River acted as a natural dam of sorts, blocking most of the water south of the San Joaquin from draining north. Several hundred miles of Sierra Nevada drained into a basin with no outlet and the result was a shallow lake ringed by thousands of square miles of marshland. The size of the lake varied by year and season, but it always abounded with fish and game. Countless millions of West Coast birds traveling the Great Pacific Flyway made Tulare Lake a biannual stopover. Indigenous Californians first reached the western lakeshore about eleven thousand years ago but didn't settle in large numbers until seven thousand years ago. Those people who did worked diligently to cultivate local food crops and develop subsistence technology, aided by the exceptional fertility and fecundity of the region. They traded and conversed in a language grouping known today as Yokuts (though their modern descendants often identify with specific tribal names). Before the European invasion of California in the 1800s, Tulare Lake and its surrounding wetlands sustained at least twenty thousand Indigenous people with a quality of life their descendants would never enjoy again.[17]

Five rivers accounted for most of the lake's inflow, and, of these, the largest by far was the Kings River—so named by Spanish explorer Gabriel Moraga (who already made an appearance in chapter 2) on his 1806 expedition. Like many California mountain rivers it is a lazy little trickle for most of the year before exploding with dangerous seasonal floods— the average flow is a little more than 2,000 cubic feet per second, with a flood peak of 90,000 cubic feet per second. Eons of water flow, both frozen and liquid, have carved out a deep and rocky canyon. Immense sequoia trees, including the famed General Grant Sequoia, colonize swaths of the valley floor. The Kings Canyon, as it's now called, stands as one of the continent's most beautiful natural landscapes. Naturalist John Muir considered it to be one of the Sierra's priceless jewels, the equal of the Yosemite and Hetch Hetchy valleys. Men who had spent their entire lives trekking through western wilderness found themselves misty and speechless at the splendor. The Yokuts peoples living along the Kings River and Tulare Lake held those places in even higher regard: they were the cradle of life itself, the grain of earth under which the waters of the primal world still ran. Yokuts mythology contained no fall from grace, no hounding from Eden. They believed they lived in paradise.

Colonization would rob them of this comfort, along with just about everything else they possessed. Cholera and malaria outbreaks in the early 1830s killed three-quarters of the Indigenous people in the San Joaquin Valley.[18] Once the United States had taken California as a war prize from Mexico, the old California that had been sheltered by law and geography from American settlers was doomed. The discovery of gold merely accelerated the process; prospectors found a few shimmering motes on the Kern River (feeding Tulare Lake) in 1853, and hydraulic miners followed to try and blast ore out of the riverbanks. When that failed, they turned to ranching. European livestock devastated the ecosystem the Yokuts relied on to survive, so the Indigenous groups became ingeniously elusive thieves of horses and cattle. Accusations of cattle theft fueled the Tule River War of 1856, which saw ranchers wage a scorched-earth campaign against a Yokuts group in the Sierra foothills. Indigenous peoples were systematically exterminated and driven from the river valleys of the southern Sierras. Ten thousand years of civilization were annihilated for the benefit of livestock.

In 1890, President Benjamin Harrison acted to protect the area's wilderness from logging and grazing claims by dedicating the General Grant National Park. Very little development occurred there, even relative to other national parks, as the park could only be accessed via several days on horseback or in a swaying, rattling stagecoach. When the land came into the control of Harold Ickes in 1933, he sensed buzzards circling. The 1920 Federal Power Act had opened federal lands to hydropower development, and California's business interests were eager to dam and log Kings Canyon.

In some ways the circumstance resembled that of the Olympic Peninsula: the U.S. Park Service tried to place Kings Canyon under its protection while the Forest Service fought to keep every acre open to exploitation. A large dam on the Kings River would not just produce copious electricity for the San Joaquin Valley; it would realize the ambitions of the state's big landowners, who had bought up land around Tulare Lake and abutting the Kings River. The soil was rich in nutrients yet poor for farming, as the bottomlands were too marshy and the river ran through a gorge so deep that its water couldn't reach the land on either side. A dam would dry out the wetlands and provide cheap irrigation water to the highlands—a perfect solution. The Forest Service agreed, and the landowner lobby was so vocal that Roosevelt's 1935 legislation to make Kings Canyon a national park died in Congress.

Harold Ickes would not be defeated so easily. Each year he listed a national park in Kings Canyon as a major goal in Interior's internal reports. In 1938 he dispatched Irving Brant to the Sierras, who reported with frustration that the controversy was "99% political," almost entirely a product of territoriality between Parks and Forestry. "Had the two bureaus been united, they could have saved both [Cedar Grove and the Tehipite Valley] by persuading the irrigationists of the truth—that better power sites existed on the North Fork of the Kings River, outside the proposed park. Instead, the Forest Service backed the demand for dams at these two points."[19] The Forest Service's regional director, S. Bevier Show, despised the New Deal, hated Roosevelt and Ickes at a personal level, and regarded their efforts as meddling in the near-total autonomy he enjoyed to give away public resources to wealthy private interests. At one point the secretary of agriculture admitted to Irving Brant that he had no control over his ostensible subordinates in the Forest Service. Roosevelt issued an executive order forbidding Forest Service employees from publicly disparaging the Kings Canyon National Park project. They went ahead and denounced it anyway.[20]

In 1939 another Kings Canyon bill attempted to summit the legislative process. This one made it through, thanks in no small part to Ickes's relentless lobbying over twenty-three consecutive days of hearings. He personally testified for three days, during which time he excoriated the corrupt business interests opposing the bill. Roosevelt signed Kings Canyon into the national park system on March 4, 1940. Some conservationists were overjoyed; others accused Interior of a long-term gambit to fill the Kings Canyon with their own dams. They were wrong on the particulars, but in a general sense their suspicions were correct. Despite Ickes's environmental bona fides, his Department of the Interior would build more dams than any agency in the world over the coming decades.

The Bureau and the Corps

Since John Wesley Powell's groundbreaking surveys of the West in the late 1860s, the U.S. government knew it was dry. For all the natural wonders Powell encountered, his overwhelming impression was one of drought to the point of desolation. The West, he spent years insisting to anyone who would listen, could never support the kind of agriculture and land use common to the nation's water-rich East. There simply was not enough rainfall to sustain large populations, not through droughts that might last fifty or one hundred years at a stretch, and trying to do so (Powell thought) was building castles on foundations of sand.[21] Yet the United States relied on westward expansion—not just for the commerce it generated but for the stability it lent to the political system. Laborers frustrated with conditions in the large eastern cities could decamp for the West, riding railroads built on speculation to attempt subsistence farming on land that just a generation before had belonged to Indigenous peoples. Workers with no stake in the American system could be handed a stake, essentially for free, so long as there remained free land to distribute. This let pressure out of a system that otherwise might have exploded, and the U.S. government had no intention of cooking its own goose. Since its founding the republic had papered over class conflict and internal contradictions by distributing free land. When the good land ran out, we invested in developing the bad land. By hook or by crook, the U.S. government would find the water it needed.

The National Reclamation Act of 1902 (referred to hereafter merely as the Reclamation Act) organized a new bureau to explore, coordinate, and fund irrigation projects in the West.

Eastern legislators rejected the bill at first until a faction of U.S. senators from the West began filibustering and otherwise blocking appropriation bills for harbor improvements and river navigation that would benefit the East. Western states were entitled to internal improvements, they insisted—just as those in the East had enjoyed generations of federally maintained waterways.[22] The Reclamation Act, establishing the U.S. Bureau of Reclamation, became law and, in 1907, the bureau was shifted to the Department of Interior. Irrigation projects were to be funded at first by sales of federal land in the western states, and subsequently by sales of water to the farmers and ranchers who bought the land. Crucially, all the money raised within a given state was to be reinvested within that state—a provision that tied the government's hands to some degree but ensured cash would continue circulating within the districts of western Congressmen. Many of those legislators had not even voted for the Reclamation Act, deriding it as foul socialism, but, once it passed, they wet their beaks as eagerly as anyone.

Of course the Reclamation Act wasn't socialist—a program designed to transfer millions of acres of land from public to private ownership could never be socialist. The Reclamation Act is better understood as an extension of the 1862 Homestead Act, which allowed citizens to claim up to 160 acres of public land essentially for free. The Reclamation Act opened new western lands to homesteading at the same limit of 160 acres for a single farmer and 320 acres for a married couple, guaranteeing them water at dirt-cheap prices and interest-free credit on purchases of both land and water. The Reclamation Act's one-size-fits-all approach met some success in the wet East but faltered badly in the geographically and climatically diverse West. As Marc Reisner writes in his classic book on western water, *Cadillac Desert*, "You could grow wealthy on 160 acres of lemons in California and starve on 160 acres of irrigated pasture in Wyoming or Montana."[23]

Few of the new settlers had any experience with irrigation farming, which is difficult under the best of conditions. It was no wonder that many farms failed, and those that survived paid almost nothing back to the government. Discouraged farmers sold their land to speculators, who used simple tricks of paperwork to amass holdings far larger than the Reclamation Act had intended.

The Bureau of Reclamation does similar work to the Army Corps of Engineers—both plan and build major water projects like dams. Both take responsibility for flood control. Where their missions differ is their interface with commerce: the bureau tasked with facilitating irrigation, the corps with maintaining shipping lanes. In practice both agencies take on a wide variety of projects, especially the corps, which followed the flow of federal dollars westward to compete directly with the bureau.

Tulare Lake became a prime example of intragovernmental struggle. Large agricultural concerns owned most of the lake itself—or rather the lakebed land that flooded most years—but the rivers feeding the lake were public property. The Bureau of Reclamation developed plans to build dams on the Kings River and the Kern River, which would render the lake's water level more predictable while securing irrigation water for small family farms in the region. Big landowners liked the notion of damming the rivers but wanted to keep all the Reclamation Act water for themselves, which would have violated the Reclamation Act. Enter the Corps of Engineers, which in 1940 submitted a competing study of the Kings and Kern. Aligning with southern California's richest and most powerful tycoons, the corps declared their intention to construct dams to control Tulare Lake's "flooding problem." Their plan would secure roughly the same volume of irrigation water as Reclamation's plan but without requiring the irrigated land be owned and worked by smallholders, and the water itself would be free. Big agriculture could own all the land in the Tulare Valley while irrigating to their hearts' content at the taxpayer's expense.

Harold Ickes fought the corps tooth and nail, begging the president and Congress for money that would let Reclamation get started first—effectively planting their flag in the Kings and Kern Rivers. He might have succeeded had the outbreak of World War II not made a precious resource of every domestic dollar. The corps took advantage of Reclamation's delay by rushing ahead with small, seemingly incidental projects on the two rivers: a diversion gate here and there, not enough to trigger Congressional hearings but sufficient to gain a decisive advantage in the inertia-based sumo wrestling of federal appropriations. The corps clung to their leverage until Roosevelt was dead and Ickes had submitted his resignation to President Harry Truman. Pine Flat Dam went up on the Kings River between 1947 and 1954; Isabella Dam has impounded the Kern River since 1953. As a consequence of the victory of the corps in this matter and the subsequent damming of the Kaweah and Tule Rivers by the corps, Tulare Lake no longer exists. The water that might sustain the lake and its natural environment—a unique and beautiful ecosystem that persisted for thousands of years—is diverted for cash crops planted by some of the world's richest agricultural concerns. Most of the old lakebed sits barren.

Agribusiness took note of the priorities of the corps and their audacity. Whereas Ickes and the Bureau of Reclamation had always been hostile, here was a government agency willing to lie, cheat, and steal on behalf of business interests—all merely for a chance to build more water projects. For the love of the game, so to speak. Business had finally found an ally in the Roosevelt administration. Ickes condemned the corps: "spoilsmen in spirit . . . working hand in glove with land monopolies. . . . [N]o more lawless or irresponsible group than the Army Corps of Engineers has ever attempted to operate in the United States either outside of or within the law."

He laid it on a little thick. Harold Ickes did not appreciate these incursions into his territory, but he was practical enough to work alongside the corps when their interests aligned. The Public Works Administration, which was separate from the Department of the Interior but still run by Ickes, shelled out many millions of dollars to the Corps of Engineers for numerous dam projects during the PWA's decade-long existence. The Bonneville Dam stoppers the Columbia River some forty miles from Portland, Oregon; the Fort Peck Dam, in the arid plains of eastern Montana, was the first major flood control project on the long, serpentine Missouri River. The Missouri would serve as the site of an odd and expensive détente between Reclamation and the corps in the 1940s, as Congress considered the agencies' competing plans for the river basin and ultimately adopted them both. That the projects, layered atop each other, didn't accomplish much irrigation *or* power generation was unimportant compared with the political benefits of pumping government money into the underdeveloped American West. Once again our system used the prospect of unlimited westward expansion to sidestep political struggle.

The Art of the Dramatic Exit

World War II introduced new imperatives to the Roosevelt administration. The New Deal's accomplishments had always been marked—watered down, left-wing Democrats insisted—by a certain pragmatism, a willingness to work with and around big business rather than directly confronting it. Committing to total war on two fronts with Germany and Japan, Roosevelt knew he'd need more industrial production than the nation had ever mustered. Such production would require large portions of the U.S. economy to operate under command of the White House. If the business community (some of whom sympathized with Hitler) refused to cooperate with federal edicts, they could have hamstrung the war effort. Roosevelt accommodated industry on more than a few issues: he opened millions of acres of federal land to mining, logging, and oil exploration. He relaxed the environmental standards of the Tennessee Valley Authority (TVA), and soon their power plants were one of the nation's largest consumers of coal. With Executive Order 9381 he enacted a nationwide freeze on wage increases for workers. The expulsion of Japanese Americans from their property on the West Coast and their subsequent internment was a land grab as much as anything else, allowing big agriculture to consolidate the vacated farms.

Harold Ickes fought many of these measures. He personally opposed the Japanese American internment, dubbing the notion "both stupid and cruel."[24] But public hysteria was so inflamed by lies from Department of War spokesmen and their credulous amplifiers in the national media that Roosevelt seems not to have hesitated at the decision.[25] The Department of War foisted management of the internment camps off to the Department of the Interior, but Ickes had no authority to return those innocents to their homes. He mostly complained in his diary and advocated for the Japanese Americans behind closed doors, unwilling to publicly rebuke the president during wartime. Roosevelt wouldn't heed his advice until December 1944, after he had secured reelection to his fourth term. Ickes also distinguished himself as a humanitarian by advocating (before the war) for the relocation of several hundred thousand Jewish refugees from Europe to Alaska, where they might build a new home for themselves. Antisemitism in the U.S. government torpedoed this exotic notion, but Ickes's plan ultimately became the basis for Michael Chabon's acclaimed alternate history novel, *The Yiddish Policemen's Union*.

Franklin Delano Roosevelt, a son of wealth and privilege who fell short of nearly all his ambitions yet still proved the best president of the twentieth century, died rather suddenly of a cerebral hemorrhage on April 12, 1945. Harold Ickes shed no tears that first night for his patron, the man who had plucked him from obscurity and given him more power than almost anyone in the government. Among the oldest New Dealers there seems to have been a sense of grim finality, that one era had perished with the next still unborn. Roosevelt had been president for twelve full years, longer than any other, and, during those years, the nation had endured unthinkable traumas: the Depression, the Dust Bowl, the horrifying war then approaching its conclusion. For a substantial chunk of the American electorate, who had come of age during these setbacks, Franklin Roosevelt was the only president they had ever known. He had won four terms handily, built the most successful and impactful political coalition in American history, and only death could unseat him. The United States had been through so much with Roosevelt as a uniquely brilliant ray of hope, and now he was gone. William White of the *New York Times* wrote, "It seemed not simply that a Man of History had died, but that history itself had lost its attribute of immortality. . . . It was a kind of grief, both interpersonal and institutional, that I had never known and would never know again."[26]

Following the president's state funeral in Washington DC, his casket rode in his customary personal train car up to Hyde Park, New York, for burial. Anna Roosevelt spent the overnight transit in her father's stateroom: "All night I sat on the foot of that berth and watched the people who had come to see the train pass by. There were little children, mothers, fathers, grandparents. They were there at eleven at night, at two in the morning, at four—at all hours during that long night."[27] Reporters along the railway that first bore the president's body from his deathbed in Georgia to Washington DC described onlooking crowds throughout the journey. Thousands of people turned out at all hours of the day and night merely to observe the passing of a train. Finally, witnessing the small funeral held in Hyde Park for the president, Harold Ickes shed the first and last public tears of his career.[28]

Vice President Harry Truman accepted the highest office in the land with apprehension and regret. Roosevelt had not kept him much in the loop—Truman wouldn't learn of the Manhattan Project's existence until two weeks into his presidency—and the task of finishing out World War II on top of governing a large grieving nation would have intimidated any man. Truman had made his name first as an army officer and then as a senatorial crusader against waste and inefficiency. He held no personal or political investment in the New Deal and Roosevelt's cabinet knew it. Despite Truman asking them to stay on, the department Secretaries began plotting their exits. Truman hated Ickes personally—calling him "shitass Ix"—and surely Ickes knew this, but the secretary of the interior was near retirement age and disinclined to make Truman's life easier by simply walking out the door.[29] Ickes bided his time; either Truman would fire him or a good excuse to quit would appear.

He didn't have long to wait. Harry Truman did some positive things like racially integrating both the U.S. armed services and all the various federal agencies, but he also displayed an alarming habit of appointing crooked officials to important posts, and his administration suffered through a number of corruption scandals. One of these involved Truman's pick for secretary of the navy, a man named Edwin Pauley who had made a fortune in the oil industry before getting into politics and who would one day involve himself with George H.W. Bush's CIA-connected Zapata Corporation. Harold Ickes hinted very strongly to Truman and his aides that he held incriminating information about Pauley's past, but remarkably nobody in the White House ever bothered to ask Ickes to elaborate. When Pauley's nomination came before the Senate, Ickes saw his opening. Before the U.S. Senate, the secretary of the interior testified that Pauley had spent the last year pressuring him to drop the federal government's claims to ocean territory off the California coast. Oil companies wanted these seabeds opened to drilling and knew they would have a much easier time winning approval from the eminently bribable state authorities than their more honest federal counterparts. To that end, Pauley (then serving as the Treasurer of the Democratic National Committee) had come to Ickes in 1944 promising $300,000 in oil industry donations to Roosevelt's reelection campaign if Ickes ordered Interior to drop their claims. Ickes, being Ickes, indignantly refused to do Pauley a single favor and told the Senate exactly what he had done.[30]

Truman had to be furious, but he blithely dismissed the allegation, saying Ickes must have been mistaken as to his recollection. This provocation tasted like ambrosia to Ickes, who lived for drama. The man Roosevelt had affectionately nicknamed Donald Duck for his stature and vocal inflection enjoyed nothing more than getting pissed off on his own behalf. He had tendered many private resignations to Roosevelt, all of them privately rejected, but this time Ickes made his intentions fully public. "This is not a tentative resignation," he announced on February 14, 1946, to an audience of newspaper reporters. "I don't care to stay in an administration where I am expected to commit perjury for the sake of a party. . . . I didn't like the President's statement that I might be mistaken about Pauley. I couldn't possibly be mistaken. . . . [Truman] should have ascertained the facts. He should have read the record."[31] When Ickes finished his oration, the assembled journalists burst into applause. He strode out of the edifice he'd personally designed into the annals of American history, loving every second of it. All Truman could do was bitch about Ickes's successful grandstanding. He withdrew Pauley's nomination.

Michael Straus and the Concrete Bonanza

An administrator like Harold Ickes leaves his influence everywhere, long after he is out the door. Upon taking his position at Interior in 1933, Ickes had appointed as his personal aide and press handler a man named Michael Straus. Straus, like Ickes, had come up through the world of newspapers, and, like Ickes, he was a prickly fellow given to argument. But the two men had grown up in very different environs: whereas Ickes came from a poor railroad town, Straus was a son of comfort who had married into real wealth through the Dodge family. He quickly earned a reputation within the department as an arrogant, self-aggrandizing man on top of being a world-class slob. But he was a committed New Dealer, an idealist who never let truth, details, or budgets impede his goals. Ickes found Straus's hard-charging advocacy useful and promoted him in 1943 to first assistant secretary of the interior. When the Bureau of Reclamation needed a new commissioner in December 1945, Ickes slotted his right-hand man into a position that had previously been occupied by buttoned-up old civil engineers like Elwood Mead and Harry Bashore. Ickes and Roosevelt meant to turn the Bureau of Reclamation into a terraforming organization, a machine to tame the West, to fill it with water and cropland and jobs and homes for the expanding American middle class.

Ickes's resignation early in 1946 left his lieutenants firmly entrenched in their offices. Truman had little appetite for further drama. The new Interior secretary, Julius Albert Krug, had come up through the TVA and supported the Department's direction. Without Ickes around to micromanage every important project, Michael Straus was left to pursue the work of remaking the West himself. He saw the region's private utility companies as his first and foremost opponent, with the corps of engineers edging out right-wing newspapermen for second place. The best way to advance the Bureau of Reclamation's irrigating mission also happened to be the best way to stick it to private utilities: build a bunch of dams. Harold Ickes saw dams as useful projects, but Michael Straus truly loved them. In Straus's eight years running Reclamation, he approved more dam projects than any commissioner before or since.

On the Colorado River the Bureau of Reclamation had built its masterwork: the Hoover Dam. The project yielded vast benefits for the people of Nevada and California, on the lower Colorado basin, in terms of irrigation, flood control, and electricity.

But it did nothing for the people of the upper Colorado basin, who looked with understandable envy at the beneficiaries of the compact. They also worried about rich, populous California using its clout to take even more water from the Colorado River than the compact laid out. To avert that frightening future they needed to use more of the Colorado themselves—put simply, they needed their own water projects.

Utah, heavily populated by Mormons, had already been pushing the limits of irrigation farming for decades. Now Colorado and Wyoming wanted in on the deal. There was one serious problem: those states had little natural capacity for agriculture. They are both very mountainous, which limits growing regions and ensures what cropland does exist lies above four thousand feet. Conditions are dry, windy, and cold—the frost-free growing season can be as short as four months. Orchards of fruit and nuts made Californian growers rich, but Rocky Mountain farmers had to support themselves by farming cheap crops like alfalfa or grazing cattle on the high plains. The upper Colorado basin was undaunted. With enough dams, reservoirs, and irrigation canals they were certain they could transform their arid territories into greenery. A fantasy, but one Michael Straus and the Bureau of Reclamation happily indulged. In order to justify the dams' expense to the rest of the country, the bureau came up with one of the twentieth-century America's goofiest yet most important concepts: the cash register dam.

Imagine a car dealership with great sales—they're the only dealer in town, everyone needs a car, and consequently they rake in cash. Now imagine the dealership also operates a mini-golf course in the back, which is a total bust because nobody's trying to buy a car and play mini-golf on the same trip. Practically speaking these are two businesses, one profitable and one unprofitable, but a creative accountant could lump their books together and leave the appearance of a single modestly successful business. This was exactly what the Bureau of Reclamation did with their dam projects: hydroelectric dams sold electricity for a large profit while distributing water to farmers essentially for free. Only one of these enterprises was profitable or efficient, but Reclamation's accounting claimed both were. Under Michael Straus such a claim was good enough to get a few million tons of concrete poured almost anywhere in the United States. The bureau first deployed this form of accounting on Wyoming's Bighorn River in 1942, and it proved invaluable for advancing their interests.

Cash register dams fed a perverse incentive for planners: the poorer a region's agricultural prospects, the bigger the dam it needed to balance the books. Americans desperate for opportunity after the Great Depression were in no position to say no. Thousands upon thousands of farmers bought up plots and claimed their 160 acres of irrigation water. They meant to create a new agricultural middle class in the high, dry West, and the federal government meant to help. But grain prices, which spiked during World War II, collapsed by nearly 50 percent once tank treads stopped ripping up the farmland of Europe. The United States found itself with surpluses of grains and beef, with the federal government feeling pressure to keep prices up through direct subsidies to farmers and ranchers.

Any conventional justification for building dams on the upper Colorado evaporated like so much reservoir water under the sun. But the people of the upper basin still wanted dams; they wanted the water they'd impound, the jobs they'd provide, the feeling of mastery that came with seeing green crops on heretofore barren plains. They wanted what the rest of the country already had. In early 1953, with Dwight Eisenhower about to be inaugurated and certain to replace him as Reclamation chief, Michael Straus proposed the Colorado River Storage Project (CRSP). A plan of alarming scale and ambition, the CRSP sought to establish nearly fifty million acre-feet of reservoirs—more than doubling the basin's storage capacity, ultimately allowing the government to hold several *years'* worth of Colorado River water in reserve against droughts. Of course, low crop prices had also ruined land values to the point where the Bureau of Reclamation would be spending up to $1,000 per acre to improve land that was not worth a tenth of that. Straus didn't care.

Paul Douglas of Illinois demolished Straus's plan on the floor of the U.S. Senate, but none of his rhetorical points counted for anything in the end—a useful reminder for everyone on the limited utility of debate in public policy. Western legislators backed the project, many hoping to get a future dam project approved for their own districts. "The [CRSP] also enjoyed overwhelming public support," Marc Reisner writes, "not just among the western farmers, but among their city brethren, too . . . Democrats, Republicans—ideology meant nothing where water was concerned." Southern Californians objected to the competition for Colorado River water, but much of the West already saw them as greedy heels. The CRSP became law, forever altering large swaths of five states and hauling Michael Straus's grandiose dreams roughly into reality. Straus loved building, loved expansion, and the final result of his tenure was a major expansion of irrigated cropland and economic development across the arid West. The New Deal expanded Americans' notions of what government could do. As the nation settled in for the 1950s, it was clear we intended to tame the West and would sacrifice any sum of nature and treasure to do it.

5

APPROACHING THE SPILLWAY

They buried the New Deal's innocence at the bottom of Glen Canyon. Seven American states signed the Colorado River Compact in 1922, and all seven felt entitled to leverage the agreement into shares of Congress's seemingly bottomless budget for western development. Hoover Dam in Nevada represented the single most important water project on the Colorado, but its benefits—jobs, electricity, flood control, and reservoir capacity—flowed entirely to California, Nevada, and Arizona on the river's lower basin. The upper basin states (New Mexico, Utah, Colorado, and Wyoming) wanted their own infrastructure to protect their water allotments under the 1922 compact. These low-population states, quite underdeveloped for the first half of the twentieth century, could not yet make use of their full water allotments and feared the lower basin states would simply take their water, appropriating more and more of the unused portion until there was none left for the upper basin to expand.

The best way to prevent that theft was a physical barrier that would keep the upper basin's water in the upper basin. Californians were opposed, but the U.S. Bureau of Reclamation took its terraforming responsibilities seriously and rarely passed up any opportunity to build a big dam on an important river. Due to the political struggle between the two halves of the river basin, the dam had to be built upriver from the town of Lee's Ferry in northern Arizona (the dividing line). As it happened, there was an excellent site not far from Lee's

Ferry: a narrow canyon with high walls of solid rock, ideal for creating the sort of artificial waterfall that drives hydroelectric turbines. The land the Bureau of Reclamation decided to flood was a remote plot surrounded by one of the world's most inhospitable deserts. Only a handful of white men had ever visited the place, let alone canoed the slumbering dragon of the Colorado River running its floor. One of those men—the legendary explorer John Wesley Powell—named it Glen Canyon.

He named it for the shaded green glens he found lining the riverbank. The Colorado River stands out among American waterways both for the unpredictable violence of its flow (see chapter 2) and the heavy sediment load it transports along the way. In Glen Canyon these traits combined to establish one of the most vibrant and valuable ecosystems in North America: the beating heart of the Colorado River's biodiversity.

Eons of moving water have carved a path through the red rocks of the Great Basin Desert. Vicious rapids cut cliffs to sheer verticals and isolated the river from the land on either side. Seasonal floods deposited heavy loads of silt, leaving sandbars exposed on the canyon floor when the floods receded. These seasonally shifting beaches played host to hundreds of plant species whose roots anchored more and more soil in place until trees could grow. Birds found these riverside oases—over three hundred species, from blue herons and Mexican owls to spider-legged canyon wrens—as did deer, antelope, and bighorn sheep that picked their ways down treacherous rock faces to graze. Game animals led the first Indigenous Americans to Glen Canyon more than ten thousand years ago, and, within two millennia, they had fully integrated the canyon's bounty into their society.

In the twentieth century, after the federal government decided to dam the Colorado River at Glen Canyon, archaeologists and anthropologists rushed out into the desert to try and document whatever they could before the canyon flooded. They found up to three thousand historic sites of Indigenous habitation: tools, knives, scrapers, and hundreds of spear-heads alongside countless woven sandals—yet garbage piles were rare and burial sites even rarer, indicating that ancient Americans didn't reside in the canyon full time. Rather they seem to have built permanent settlements in the craggy mountains of the high desert many miles away. A picture emerges of a civilization where children grew up in mountain warrens under the watchful eyes of the elderly, as their young workers made the annual hike to Glen Canyon. During the summer and fall months, these early Americans used the canyon's narrow confines to corral and slaughter dozens of game animals, dressing and preserving the meat while others gathered vegetables from the natural gardens along the river. At season's end they hauled what they gathered out of the canyon and back to the high desert where the cool dry air could preserve it for the coming year.

By about 400 BCE the people of the region had developed a more deliberate style of agriculture, planting corn and squash along the canyon floor once each year's floods had finished rearranging the landscape. Populations from both the Fremont and Ancestral Puebloan cultures lived in Glen Canyon from 500 CE until roughly 1300 CE, living in permanent settlements carved into the red canyon walls and feeding themselves almost entirely through agriculture. But by the year 1500 they were migrating away from Glen Canyon, driven perhaps by the historic megadrought that afflicted the Southwest during that era. The Navajo, Hopi, and Paiute nations all walked the canyon, and their descendants attach great spiritual value to many locations within, such as the three-hundred-foot free-standing arch of the Rainbow Bridge.

Interior Secretary Harold Ickes tried to preserve all of Glen Canyon and much of the surrounding region on the Colorado and Green Rivers. He proposed the Escalante National Monument in 1936 as part of his campaign to expand the national park system (see chapter 4) but ran into resistance from Utah legislators whose constituents wanted unlimited rights to run their cattle over federally owned land. World War II ended Ickes's ambitions for the park, especially once prospectors discovered uranium in the area. Every incentive ran toward building a major dam for the upper basin and, by 1956, the fate of Glen Canyon was already decided. Many accounts of the Glen Canyon Dam negotiations blame the environmental movement for its own defeat on this front: they are accused of allowing the Colorado River to be dammed in exchange for the cancellation of another planned dam on the Green River. This Faustian bargain, the story goes, traded a dam for a dam and cost the West one of its great jewels. Sierra Club president David Brower, stricken by guilt, contributed to a book called *The Place No One Knew* by way of apology to the nation.

Brower's guilt was misplaced. As detailed earlier, the Glen Canyon Dam was a perfect sop to the concerns of upper basin landowners and the legislators they funded. The Bureau of Reclamation clamored for the project; the bureau's leadership saw Glen Canyon as a top priority before the dam went up and still consider it one of their all-time greatest success stories. Put simply, Glen Canyon Dam was always going up, if the feds had anything to say about it. Brower, despite his pulpit atop the influential Sierra Club, never traded a dam for a dam.[1] The U.S. government would never give an environmentalist that kind of power, certainly not at the apex of America's dambuilding fervor, with the Bureau of Reclamation nearly as powerful as it would ever be.

A rush to the desert followed the dam announcement. Nature lovers, both professional and amateur, drove their cars for hundreds of miles over gravel roads and stretches of sandstone polished so smooth that tires could find no purchase. Locals who had long ago fallen in love with this gorgeous and secret place trouped back together for a few last hurrahs before it would all be taken away. Their photographs depict lush greenery on the canyon's sandbars—not desert shrubs but real trees, stands of gnarled and durable oak backed up by hanging gardens high on the walls where the water seeped through Navajo sandstone. Narrow side canyons formed and, fed only once a decade by rare storm surges, enticed hikers down secluded passageways hundreds of feet deep yet only wide enough to walk single file. Ferns and orchids lined the dripping walls. Every side canyon was its own adventure, totally unique among hundreds. Iceberg Canyon held a deep pool of still water so cold it took a strong swimmer to cross without her muscles seizing. Dungeon Canyon had high inverted walls that blocked almost all sunlight and beamed the remainder into deep and dramatic shadows, creating a forbidding grotto like the depths of a medieval dungeon. Mystery Canyon was named for a set of Ancestral Puebloan–cut handholds in a steep rock wall that simply stopped at an impassable face. Twilight Canyon was a semicircular natural amphitheater facing west, where generations of Indigenous people had carved pictographs of humans, animals, supernatural beings, and intricate geometric patterns, all to greet the fire of the setting sun. In the evenings the whole place was ablaze in orange as the sky slowly bruised.

These were the lesser attractions, Glen Canyon's deep cuts. Marquee features included the Tapestry Wall, a smooth vertical plane rising one thousand feet above the natural river bed. Water filters down through sandstone until it reaches a less permeable layer of rock, at which point it flows sideways and out into free air. Seeping liquid dribbles down the pale sandstone face, staining it dark like artfully streaked paint, creating a wholly natural mural hundreds of meters across.

Cathedral in the Desert hides in an unassuming tributary canyon off the Escalante River, which flows into the Colorado some miles above the Glen Canyon Dam. Before the river's flow was stopped, many considered the Cathedral to be one of the most arrestingly beautiful natural spots on Earth. Eons of running water have carved a stunning chamber from the soft sandstone. Hidden from the bright sun and rushing river, the Cathedral's high walls of yellow and orange trap light and sound for a sensation of perfect serenity broken only by the mantra-like drumming of a sixty-foot waterfall dropping into a deep pool ringed by green moss. Photographer Tad Nichols, who shot annually in the old Glen Canyon, wrote in 1999, "You can't do justice to that place by talking, even by pictures. You had to stand there and see it and feel it. I'm sorry most people couldn't do that."[2]

Twenty years of drought in the West have proved Nichols wrong. The water level in Lake Powell, the artificial inland sea accumulated behind the Glen Canyon Dam, has lately dropped lower than at any other point since the reservoir first filled. In recent dry seasons the Cathedral's waterfall has consistently been exposed. If the water's elevation drops below 3,557 feet, the Cathedral will have a floor once again and, if the dam is ever removed, this magical place will be one of the first to emerge from Lake Powell's tarpaulin blue.

Some venues are lost more completely. Music Temple, first mentioned in John Wesley Powell's famous journal, is a domed hollow that, before the dam, had a single entrance hidden in greenery on a sandbar beach:

> We find ourselves in a vast chamber, carved out of the rock. At the upper end there is a clear, deep pool of water, bordered with verdure. . . . The chamber is more than 200 feet high, 500 feet long, and 200 feet wide. Through the ceiling, and on through the rocks for a thousand feet above, there is a narrow, winding skylight; and this is all carved out by a narrow stream which runs only during the few showers that fall now and again in this arid country. . . . When "Old Shady" sings us a song at night, we are pleased to find that this hollow in the rock is filled with sweet sounds. It was doubtless made for an academy of music by its storm-born architect; so we name it Music Temple.[3]

Approaching the Spillway

Twentieth century visitors to Glen Canyon would gather at night in the Temple to play instruments and sing to the reverberating darkness, but no one has set foot in that special place since the dam went up. Many visitors photographed a flat chunk of orange sandstone on which Powell and other nineteenth-century explorers had carved their weighty names. On his last visit he happened to arrive shortly after a rainstorm, and, for the first and only time, he captured an active waterfall roaring through the Temple. Today that cool, mysterious entry passage lies at the bottom of the lake. It is hard to imagine anyone seeing it again within our lifetimes.

Flooding Glen Canyon destroyed so many natural relics—places that arrested the breath of seasoned outdoorsmen, places Indigenous Americans believed were walked by the gods themselves—yet the loss of the mundane stings nearly as much. Tad Nichols writes, "At the base of this cliff wall near Hall's Crossing was a grove of oak trees. At the top of the oaks was a great blue heron colony. There were always herons there, and we could see their nests . . . as the lake came up, those trees were flooded. But the herons came back to the same spot and tried to nest on the wall. Even though the trees were gone, they came back to the same place."[4] For generations untold those huge birds rooted their lives at one spot. What must they have thought, or felt? Glen Canyon took millions of years to form and with the turn of a screw we took it all away. To a heron, we might as well have untethered the sky.

Floyd Dominy and the Vacationer's Paradise

If rivers are dragons, then Floyd Dominy was the greatest dragonslayer in American history. Western waterways sharpened their claws on the steep course down from the Rockies, and Dominy, as the Commissioner of the U.S. Bureau of Reclamation, meant to bring them all to heel that American civilization might flourish. Born in 1909, he joined Reclamation in 1946 and spent most of the 1950s supervising outlays and revenues for the bureau's Division of Irrigation. Dominy possessed a penetrating intelligence unburdened by ideology or morality—he perceived the world in an unvarnished fashion and believed in nothing above or below. Answering to dryland farmers indignant that a proposed Reclamation project might force them into irrigation farming with a deluge of free water, Dominy called the project "damn foolishness" but laid out the political rationale: "The local towns and businessmen wanted it. They could see themselves growing fat on large-scale construction payrolls. They could see something to be gained by increasing the number of farm families in their service area. Like the usual selfish citizen, they were willing to accept this increase to their personal larder without thought as to the burden to be placed on the Federal taxpayer."

As one might surmise from that passage, Dominy was an unpleasant man with a great deal of scorn for other people and a dim view of human nature. He was also one of the most powerful and effective federal bureaucrats of the twentieth century, especially among those who never reached the president's Cabinet. His upbringing on a small ranch in southern Nebraska and his professional background in water accounting made Dominy acutely aware of what water could do. More importantly he understood what it meant to people, especially to the rural yeomen of the West. To those people water held value far beyond its market price—it represented the very viability of their communities. It represented the future, the American Dream they'd all been implicitly promised.

As Dominy once told a Congressional hearing regarding the abysmal repayment rates on Reclamation water contracts, "I told them, I said, 'This project needs to have not 40 years' repayment period, but 140 years.' The Congressman said, 'Well, this looks like infinity.' I said, 'That's right, it will never pay out.' 'Well then, why don't we abandon it?' 'Well, we don't abandon it because we have a viable community out there. We've got a small city, we've got a viable environment if you let those settlers pay what they can afford to pay, which is very little."[5] Western expansion meant cheap land, water, and homes heavily subsidized by the government, and the Bureau of Reclamation existed to facilitate those subsidies. Floyd Dominy understood it all perfectly. Yet he wasn't a total cynic, once telling environmentalist John McPhee of the dry plains of Wyoming, "I watched the people there—I mean good folk, industrious, hard-working, frugal—compete with the rigors of nature against hopeless odds. They would ruin their health and still fail."[6]

Dominy's efficiency and determination got him a position in Washington DC as head of the Division of Irrigation. In hearings he served as a "backstop," seated immediately behind the Bureau of Reclamation's assistant commissioners where he could feed them detailed answers to Congressional queries. That role expanded in 1954, when Congress reassigned Reclamation's duties from the Interior and Insular Affairs Committee, on which many Western legislators sat, to the Public Works Committee, where nobody hailed from the dry West. Dominy realized the bureau's disadvantage and pleaded with his superiors for a tactical change: "We can't just answer the questions, we've got to make a speech, we've got to sell Reclamation. This group is not friendly . . . they want all the money to go to the Corps of Engineers and the Eastern part of the United States."[7]

The commissioners ignored his advice. The resulting hearings were so painfully antagonistic and unproductive that, in 1957, the secretary of the interior (Fred Seaton, also from Nebraska) promoted Dominy to associate commissioner, "and [Bureau of Reclamation] Commissioner Dexheimer was told to stay out of my way, and I really ran the Bureau for two years before I was made Commissioner."[8] This sounds like some bullshit a person might say to talk himself up, but Marc Reisner's book, *Cadillac Desert*, confirms Dominy's power play. The year

1959 saw Dominy's formal elevation to Commissioner. The bureau was occupied with dozens of dam projects, including several major works already approved or under construction: Trinity Dam in northern California and Yellowtail Dam on Montana's Bighorn River were two of the largest. River basin accounting let dams cover massive financial losses on irrigation water by balancing the books with hydropower sales, thus the bureau had every incentive to build bigger and taller. Private power utilities across the West complained bitterly, but the ideologically uninhibited Dominy made peace where early New Dealers like Harold Ickes would have twisted the knife. He cut deals with the utilities, selling cheap hydropower to them directly so they could turn around and sell it to their customers at higher rates. Dominy didn't mind as long as the dams went up. Decades later he reminisced fondly of these years, "This is the stage I like to describe as having no naysayers."[9]

The Colorado River proved trickier. Sierra Club activists led by David Brower fought to protect what they could. They managed to get the Echo Park Dam canceled yet saw Glen Canyon drown starting in 1964. Filling Lake Powell, the Glen Canyon Dam's reservoir, with water equivalent to two years of Colorado River flow took nearly a decade. That slow death only made the loss more painful, and American environmentalists increasingly began to believe that the hundreds of giant dams going up all over the country were each an environmental disaster strangling natural waterways.

Floyd Dominy saw rivers in a very different light. "I don't think we destroyed the Gunnison River," he declared, speaking of several much-protested sites in the state of Colorado. "Before, it was a closed river. Now the public has access to those three reservoirs, with far more fishing and boating and recreational activities than were ever there in its natural state."[10] Dominy loved the notion of tourism as an economic catalyst. He spent decades extolling the virtues of "recreational activities" around reservoirs, especially those activities newly available on Lake Powell behind the Glen Canyon Dam. "Rainbow Bridge gets 300,000 [visitors] or more a year now, when it had only 15,000 in 50 years. The Colorado River float trip was limited to about a six-week period, haphazardly, when it was available in the flood season. Now you've got 20,000 people a year going down it."[11]

The dam turned Glen Canyon from a richly textured natural environment into an immense and somewhat eerie artificial lake. Damming a fast, sediment-rich river like the Colorado works a fundamental change on its environment almost immediately because the water slows to a crawl. Tributary rivers carry sediment to the Colorado, which is supposed to flush it all downhill toward the ocean. Instead that sediment runs into tranquil Lake Powell, where without propulsion it merely sinks. What particular matter does reach the Glen Canyon Dam cannot pass through its penstocks and builds up at its base. As a consequence the water downstream from the dam carries only a small fraction of the nutrients it used to, barring the occasional "pulse" of water and sediment released by the bureau through the dam. Yet those nutrients don't circulate within the Lake Powell ecosystem, either. They just build up, allowing heavy metals to concentrate and become toxic. The water in Lake Powell is not just unusually still and clear, it is unusually cold as well: 44–53 degrees Fahrenheit year-round, as opposed to the naturally flowing Colorado, which rumbled with ice floes in the winter and could get as warm as eighty degrees in late summer.[12] Rivers equilibrate their water temperatures pretty well as they run, since the sun can warm the riverbed, but in deep lakes the top layer of water insulates the bottom from heat and light. The lower levels become a dark and frigid tomb while fish and other animals compete in the upper levels for what scarce nutrients remain.

Visually Lake Powell is quite striking, like an artist's conception of water on Mars. Sheer cliffs of yellow sandstone rise from the sapphire-blue lake, without a scrap of vegetation to break up the solid lines and smooth textures. High canyon walls and colossal buttes are a constant reminder you're on a river, though the lake has no discernible current. No large animals move along the shore, for there is nothing to graze, though lizards, snakes, and rabbits can be found down side canyons. Rarely do birds cross the sky, not even in the morning and evening when one would expect them to feed, and, during these times, the rocks glow fiery orange while the lake stands utterly still and quiet. The serenity is special, one of the major reasons tourists flock to Lake Powell, yet it is artificial—Glen Canyon's native bird populations are permanently diminished, along with its native fish. What used to be a warm, shallow, fast, high-sediment environment has become the polar opposite on all counts: cold, deep, slow, and clean. Glen Canyon's native species are specialists and they have suffered badly in Lake Powell. A few have gone extinct, and the remainders are largely banished to small tributary rivers that remain undammed.

The Bureau of Reclamation works for the welfare of human beings, not birds or fish. In fact, animals like the Colorado pikeminnow are undesirable to men like Floyd Dominy—as large, aggressive predators they undermine the populations of introduced trout that Lake Powell visitors love to recreationally fish. Better to extirpate them, to make way for replacements with more tourist appeal, for the volume of dollars at play is larger than anyone who hasn't seen the shining white houseboats packing the channel at Antelope Point Marina. These floating plus-size RVs rent for several thousand dollars per week—considerably more if you pick a luxury model—and that does not include insurance, fuel, additional vehicles (like smaller boats or jet skis) and food for the large crew you've probably brought along. Camping on an enormous beautiful tranquil lake in total comfort with every modern amenity is just as nice as it sounds. Lake Powell's clientele tend to be fairly wealthy families, given the cost of enjoying the lake.

Vindicating Floyd Dominy's vision, Lake Powell is one of the most popular vacation spots in the United States. Few assets in the federal portfolio can match its appeal. In 2019, the last year before the COVID pandemic throttled domestic tourism, Lake Powell hosted 4.4 million visitors, who directly spent $427 million in various communities adjacent to Glen Canyon. The same year, Yosemite National Park also attracted 4.4 million visitors, and Yellowstone National Park attracted 4 million.[13] Lake Powell has come to represent an irreplaceable economic asset to the communities of the Arizona-Utah border region, all made possible by the accounting fiction that is the Glen Canyon Dam. The cash register did exactly what it was meant to do: inspire investment where none existed, develop barren land into productive new real estate.

That arrangement makes it difficult to imagine the region accepting major alterations to Lake Powell's status quo. Since the lake filled, conservationists have imagined various scenarios that might bring down the dam and liberate the canyon. In this they have ironically been aided by the droughts that have afflicted the West for the last two decades. Each dry year that drains Lake Powell leaves the water level lower, meaning less hydropower head for the Glen Canyon Dam. The hydropower of a dam goes up by the square of its head, rather than linearly, so even modest gains and losses of head have big impacts on a dam's output. Meanwhile, Lake Powell actually compounds the drought situation by losing a great deal of water to evaporation. A huge lake under a punishing desert sun will naturally tend to dry up, so Lake Powell loses an average of *125 billion gallons* every single year. Water seeping through the lake's expanded footprint into the surrounding sandstone swells that figure further. Making matters worse, just a few hundred miles downstream is an even larger reservoir: Lake Mead, behind the Hoover Dam, which itself gives up at least 200 billion gallons per year.[14] The two largest reservoirs on the Colorado River lose enough evaporated water to supply the entire city of Los Angeles. The entire state of Nevada does not use what these reservoirs waste every year. The two reservoirs represent the leakiest faucets imaginable, and both their hydroelectric dams, their raisons d'être, are underperforming as a result. The enormous infrastructure we have built to tame the Colorado and store and deliver its water struggles to function in the megadrought that has defined life in the American West since the year 2000 (see chapter 10).

A Body of Silt

Under Floyd Dominy's leadership the Bureau of Reclamation built hundreds of dams and planned many more that never broke ground. The Army Corps of Engineers added hundreds more. The vast majority of these projects were nowhere near as politically or economically important as the Glen Canyon Dam. As the 1950s ended, a swelling proportion of dams turned out to be little more than grist for the federal construction mill. Even as the value proposition for big dams waned, the government kept putting them up because they were necessary to develop the West.

The western states, eager to receive their own shares of the expansionary bonanza, began to plan their own dams and reservoirs. California, appropriately, had the grandest ambitions. Occupying a huge swath of the West Coast from the Sierra Nevada to the sea, it boasts some of the nation's richest soil and huge amounts of water—but rarely in the same place. Instead most of the state's water circulates in its forested northern climes (mountainous, with poor soil) while the extremely fertile southern half sits parched. This state of affairs demanded remedy. Californians dreamed up an infrastructure network of a scale and cohesion unprecedented outside the Tennessee Valley, funded by the federal government but controlled by state officials along with the large agricultural interests that kept those officials in office. It would right nature's imbalances, ending floods and droughts in one fell swoop while guaranteeing prosperity for the future. They called it the Central Valley Project (CVP).

Early in the twentieth century the California legislature approved a major dam high on the Sacramento River, near the Oregon border, but could not sell enough bonds to start construction. Roosevelt's presidency and Harold Ickes's tenure at the Department of Interior opened new doors: the Bureau of Reclamation agreed to take on the stricken dam project in 1935 as part of Roosevelt's scheme to invest in infrastructure. To lead the effort they chose Frank Crowe, chief engineer of the Hoover Dam and a living legend. Crowe deployed his signature cables-and-towers method (see chapter 2), finishing the Shasta Dam two full years ahead of schedule in 1945.

Shasta Dam would become the linchpin of the entire state's irrigation system, similar to the role the Wilson Dam at Muscle Shoals, Alabama, played for the nascent Tennessee Valley Authority (see chapter 3). It was not a "cash register" dam, which subsidizes unprofitable development with profitable hydropower, but rather a kind of central bank for Californian irrigation. The Friant Dam, built in the foothills above Fresno on roughly the same time frame as Shasta, did for the smaller San Joaquin River what Shasta did for the mighty Sacramento. Both had to be diverted if the Central Valley they jointly fed was to be properly developed. Most people around the planet associate California with the technology and entertainment industries, but agriculture has always been the state's most consistent economic engine. In 2019 California's seventy thousand farms and ranches earned $50 billion in revenue, collecting more than 13 percent of the United States' agricultural receipts, and almost none of those businesses drew their water from a naturally flowing channel.[15] It is all irrigation and groundwater pumping, neither of which the Californian economy could presently survive without.

Private electrical utilities pitched a fit at the state's new dams. The flood of cheap hydropower they produced threatened to fatally undermine the low-yield-high-cost model that made utility owners rich. Leading the resistance was the Pacific Gas and Electric Company (PG&E), antagonist to generations of Californians and holder of a virtual monopoly on electrical generation and distribution in the Central Valley. PG&E could not stop the dams from being built, so they challenged the Bureau of Reclamation's right to build powerlines and other infrastructure that might compete with theirs. In the end, Harold Ickes and the Department of the Interior resolved the conflict by awarding PG&E contracts to distribute the electricity Shasta produced; the government got guaranteed payouts, and private utilities got to keep raking in profit. So what if Californians had to pay higher rates for the power their tax dollars were already paying to generate? So what if this arrangement locked PG&E into its monopoly position, entrusting much of the state's electrical infrastructure to a private company with every incentive to underinvest in maintenance? Ickes hated private utilities more than almost anyone in government, but the Roosevelt administration had the Second World War to win and badly wanted peace with Corporate America. They were willing to compromise, even if compromise meant locking their political opponents into power long after the war ended.

Over the following generations PG&E would use their near-monopoly in California's electrical market to systematically skimp on maintenance, leading to hundreds of preventable fires across the state and dozens of deaths. Those fire-based liabilities totaled roughly $30 billion, enough to force the company into bankruptcy protection in 2019. Yet for all the company's sins, it still wielded sufficient clout to dissuade California's Democratic-led government from attempting to take control, and, in 2020, emerged from bankruptcy nearly unaltered—subject to just a few new regulations that may ultimately prove toothless. In allowing PG&E to evade structural consequences for such historic malfeasance, California may have missed a truly rare opportunity.

North to South

Unsatisfied merely with mastery of the Sacramento River, Californians moved to take control of the other waterways in northern California. If one is determined to augment agriculture in a state, and the primary factor limiting agriculture is water scarcity, then any river or stream that reaches the ocean is wasted. California sought to transform literal streams into income streams for the CVP's water bank. That bank would direct water where growers needed it to go, much the way a traditional bank finances development. The difference between the CVP and other irrigation projects was that instead of creating new farmland, the CVP directed resources to farms that already existed.

Bailing out hardscrabble farmers whose wells had run dry would have been a defensible policy, but large interests, including railroad and oil companies, had bought most of the land years before. Getting cheap water from taxpayer-funded infrastructure sounded like a great deal to them. Reclamation law stated only small landholders (no more than 160 acres for one person and 320 for a married couple) could receive subsidized water, but some companies evaded that requirement with accounting chicanery while many others preferred to lock up any enforcement through endless legal challenges. While speaking to journalists, these landowners excoriated the government as land-grabbing tyrants. Privately they considered the Bureau of Reclamation entirely toothless. Even the pugnacious Floyd Dominy never challenged the people who were scamming his department because he saw the whole operation as a kind of benevolent scam. When lists of reclamation law violators landed on his desk, he made sure they disappeared.[16]

One by one, CVP dams stopped up the rivers of the Sierra Nevada. A wave of Biblical floods in the winter of 1955 killed more than twenty Californians and greased the skids politically. The Bureau of Reclamation raised Trinity Dam on the Trinity River (built 1957–63) and the Oroville Dam on the Feather River (built 1961–68), establishing the second- and third-biggest reservoirs in the state after Shasta. They badly wanted to dam and control the wild Klamath River as well, which draws its headwaters from the southern range of the Cascade Mountains in Oregon. The Klamath dips south into California for a portion of its run, and the bureau devised a plan to ambush it there, damming the river and using the dam's hydropower to reverse the flow. They meant to pump it uphill, sixty miles through an aqueduct, before joining it with the CVP's central water bank.[17] In the end this plan proved too audacious even for a government that had grown accustomed to scoffing at nature and geography. The lower Klamath remains one of the longest stretches of natural river yet to be found in the West, and its gravel beds spawn most of the salmon caught in the state of California, but irrigation dams along many of its important tributaries have reduced the volume of water that reaches the Klamath proper. The Klamath's salmon runs and the Indigenous peoples who rely on them have suffered terribly as a result (see chapter 10).

Once California had its water bank up and running, it needed to transport the water south to growers and ranchers. The Sacramento River represented an easy natural pipeline for half the distance, traveling due south at a shallow slope, but once water reached the Sacramento delta, gravity wanted to carry it toward the ocean. Delivering water to the southern Central Valley meant moving it uphill. From the Clifton Court Forebay, just one meter above sea level, the CVP uses powerful electric pumps to kick an artificial river into reverse-motion. The pumps accelerate water up a steep climb, after which it coasts smoothly down a shallow gradient to the next pumping station. The California Aqueduct, named for former California governor Edmund G. Brown, can be easily tracked as you drive on the U.S. I-5 highway. A tranquil body clad in cement, the aqueduct pipes billions of gallons south as smaller canals and ditches sip fluid away to supply mile upon mile of orchards and vineyards. Rich agriculture stretches all the way to the hills in the distance, at which point it abruptly stops in favor of the dry yellow scrub that would naturally carpet the region. The aqueduct marches by under the towering presence of the San Luis Dam (built from 1961–68), which serves as the rare example of a dam without a river. California created the San Luis Reservoir out of whole cloth, inundating fertile bottom-lands used by generations of Indigenous Yokuts people with flows piped in from elsewhere in the state. In a sense the San Luis Reservoir serves as a local branch of the state's water bank, storing excess winter flows closer to the growers and ranchers who have always been the CVP's first beneficiaries.

South of the San Luis Dam the aqueduct bifurcates. One path carries water to supply the perennially parched Central Coast, from San Luis Obispo down to Santa Barbara—a gorgeous stretch of country kept superficially green by ocean condensation but blessed with almost no rivers. The other runs south to the foot of the Tehachapi Mountains, just north of Los Angeles. If you drive along that stretch of I-5, as millions of people do each year, you may notice a large facility at the base of the Tehachapi foothills with a number of mammoth pipelines running uphill away from it. This is the Edmonston Pumping Plant, one of the largest such facilities in the world, where geologic weights of water are sped up and launched over a three-thousand-foot mountain range. The Tehachapis are too seismically active to tunnel beneath, and Los Angeles is too rich and thirsty to go without water, so over the Grape-vine (what everyone in California calls that stretch of high-way) the water must go.

Back in the 1950s and 1960s some citizens, politicians, and economists worried these systems might turn out to be imprac-tical boondoggles. A. D. Edmonston, California's chief water engineer, answered these objections by stating the market would simply provide: "It is indicated that urban communities will always be able and willing to pay the cost of water to meet their municipal needs. . . . [U]nder pressure of future demands for agricultural produce, the water necessary for greatly expanded irrigation development will be provided, at whatever cost may be required." One can identify this precise flavor of ideology in twenty-first century claims that wealthy cities like Miami will always be able to pay for flood mitigation, no matter the cost. While the prediction seems savvy from a certain historical perspective, the state of affairs it describes must logically have some terminus. Inevitability has a way of sneaking up on us.

The Creaking of Flashboards

By the turn of the 1960s the New Deal stood unchallenged in mainstream America. Franklin Roosevelt remained a legendary figure to tens of millions who had endured the Depression years only to witness victory in World War II and two decades of unprecedented economic expansion at home. That generation's children, then emerging from adolescence, had grown up in the era of universal electrification, interstate highways, and summers boating on the local reservoir. Very few had experienced a major flood. It would certainly be false to characterize the United States of the 1950s and 1960s as a paradise, especially with regard to anyone who was not white, but Americans of this era were more secure in many ways than ever before. Most could look to the hardships their parents faced and see their circumstances had clearly improved. Both my grandfathers went to college only because of the GI bill, after which both took up unionized white-collar professions (a draftsman and a schoolteacher, respectively) they could not have accessed without that education. In a single generation the policies of the New Deal elevated them to the middle class, and both raised their children as Roosevelt Democrats. The American system left people behind, too, but to many it delivered on its stated objectives of peace and prosperity. American products were the best in the world, our culture the most desirable, our economy a limitless engine of growth.

Roosevelt's Democratic Party reaped the rewards, dominating domestic politics for decades. Republicans controlled Congress from the 1918 elections until Roosevelt's Democratic wave in 1932, from which point Democrats ran the legislature nearly uninterrupted until 1980. Between 1933 and 1969, only one Republican won the presidency: Dwight Eisenhower, a moderate who supported the New Deal for the straightforward reason that provisions like social security were very popular, and dismantling them would inevitably prove unpopular. Eisenhower loved infrastructure projects: under his leadership Congress approved both the interstate highway system and the St. Lawrence Seaway, connecting the Great Lakes to the Atlantic Ocean via a chain of ship locks to climb hills and canals to bypass rapids. The far right wing of the Republican Party always opposed the New Deal, but these figures enjoyed little public support and typically phrased their objections in anti-Soviet terms. All of which is to say that by the time my parents were born in the mid-1950s, Americans had developed a rough consensus regarding what national prosperity looked like and how it might be sustained: consistently rising wages and expanding opportunities for average people, subsidized by ambitious government investments in infrastructure, healthcare, and income support. That was the New Deal formula.

It worked, and people liked it. Roosevelt's antagonists—finance, large corporations, and landowners—never went away or changed their minds, but the program's popularity forced them to mount their attacks from difficult angles. They accused New Dealers of being communists or at least furthering the Soviet Union's interests; they protested that the Tennessee Valley Authority or the Shasta Dam deprived private utilities of the profits they were owed in perpetuity. In truth, the right wing did not want the government spending so much money on people at the bottom of the income distribution. It was a losing argument. Few Americans wanted to beggar their neighbors in times of such plenty, and besides most of them had grown up watching the government flail helplessly during the Depression. They knew firsthand the fallout of failed governance and were not going back.

So when it came to water and infrastructure, Democrats and Republicans governed much alike. A bureaucrat like Floyd Dominy could come up under Harold Ickes and Franklin Roosevelt, take over the Bureau of Reclamation under Eisenhower, and continue to serve in that post through Presidents Kennedy and Johnson—a decade at the bureau's helm, more than three thousand days spent bullying Congressional staffers for money to remake the American West. During the go-go period from the end of World War II deep into the 1960s, the Bureau of Reclamation and Army Corps of Engineers put up thousands of dams. Their choices would impact Americans' lives for centuries, yet the deeper they advanced their schemes, the less sense they seemed to make. Increasingly their projects came to reflect the rough and unprincipled dealings of men like Dominy. Simultaneously, the other organs of the federal corpus were growing more comfortable with dubious tactics of their own.

6

GREAT, STILL MASS

If the New Deal era meant decades of swelling prosperity to citizens of the United States, to those abroad it meant the madness of the Cold War. V. I. Lenin's successful revolution in 1917 posed an existential question to societies around the world: What if there existed a genuine alternative to industrial capitalism? What if the rules and patterns that govern our lives were merely the dictates of men and could thus be rewritten? One did not need to consider Lenin a saint or the Soviet Union a model society to appreciate the power latent in that idea. Fearing that power, President Woodrow Wilson stood with the industrialized capitalist nation-states of Europe in refusing to diplomatically recognize the Soviet Union.

The United States remained in that stance, the State Department's head planted firmly in sand until 1933, when Franklin Delano Roosevelt made rapprochement with the Soviet Union one of his first international priorities. In November of 1933 the United States recognized the Soviet Union, and normal diplomacy commenced. Roosevelt was no communist nor even a socialist; he hoped the Soviets would counterbalance the empire of Japan's expansionary impulses in East Asia.[1] It was time to acknowledge reality. Roosevelt's initiative paid off, obviously, when the Soviet Union allied with the United States to win World War II. But the relationship ended with the war. The victors rushed to divide central Europe, the Soviets developed an atomic arsenal to deter the United States from leveraging *our* atomic weapons internationally, and, over the next few decades, this frustrated (and frustrating) rivalry would impact nearly every person on the planet.

Opposing the Soviet Union and its proxy states around the globe became part of the American political consensus, as unquestioned as dambuilding or social security. Presidents Truman and Eisenhower were of different political parties but shared military backgrounds: the former reinstated large numbers of Nazi officers and officials to rebuff Soviet influence in West Germany, and the latter launched the Korean War, which killed at least 10 percent of that country's population. Eisenhower also approved clandestine and military measures to support the brutal dictator of Cuba, Fulgencio Batista, for the reason that Batista was friendly to U.S. business interests. When Batista's regime fell to Fidel Castro's revolutionary uprising, the State Department worked tirelessly with the Central Intelligence Agency (CIA) and Pentagon to undermine Castro. When the United States attempted to sabotage the Cuban economy, Castro responded by seizing the assets of U.S. companies and pursuing stronger ties with the Soviet Union. In its waning years the Eisenhower administration even formulated a plan to invade the island of Cuba, overthrow Castro, and install pliant leadership. The CIA trained a small army of exiled Batista supporters to do the legwork.

Because imperialism was a consensus position, handing over the White House to John F. Kennedy changed very little. In fact, the new president was so eager to demonstrate his anticommunist credentials that, when informed of the Cuban invasion plans, he accelerated the timetable. But Kennedy wanted to have his cake and eat it too: with the aim of keeping the American involvement deniable, he insisted that no official U.S. military assets be used to support the invasion. The CIA's plan relied on both naval and air support, but they moved ahead anyway knowing that, if they called off the invasion, Kennedy might never give them another chance. On April 17, 1961, just three months after Kennedy's inauguration, the

CIA made their move. Their stateless army, twelve hundred in number, sailed from Nicaragua to the south coast of Cuba, where they attempted to land at a narrow knife-shaped inlet named the Bay of Pigs.

Two days later the invaders had been completely surrounded and defeated, the vast majority captured by Castro's soldiers. U.S. negotiators spent more than a year negotiating ransom. Castro humiliated a superpower, and, instead of returning Cuba to American influence, the operation shoved the island nation away permanently.

Kennedy's hawkish-yet-bumbling administration managed to consistently undermine the nation's image abroad. Just a month after the Bay of Pigs fiasco the undaunted Kennedy approved combat missions for U.S. aircraft into South Vietnam. By the end of 1961 he had increased American military personnel in the country to over three thousand soldiers. Like the Cuban situation, Kennedy inherited a Cold War project from the prior administration (in this case an attempt to keep France from losing their colonies in Southeast Asia to insurgency) and felt compelled by American political consensus to escalate the conflict. By the time Kennedy perished to an assassin's bullets in 1963, sixteen thousand American troops were in Vietnam.

Citizens on the left flank of the political spectrum objected to these interventions from the start, but their voices were ignored as morally suspect or possibly even pro-Soviet as the USSR made no secret of arming the forces of North Vietnam. Nonetheless these people grasped a fundamental contradiction then emerging in the New Deal's heterogeneous and vast coalition. If the U.S. government truly supported working people, if it truly believed in democracy and equality under the law, then how could it lend material support to corrupt dictators like Batista or colonial killers like the French in Vietnam? How could it conscript its own citizens to fight wars of choice around the globe? To the average American in 1961 these were questions for philosophers. A decade hence they would inflict a mortal wound upon the New Deal Democrats.

Too Much and Not Enough

The most important question in domestic policy was how best to sustain the prosperity Americans had enjoyed for half a generation. Evidence had lately emerged that Roosevelt's long status quo might need updating: recessions in 1958 and 1960 temporarily drove the unemployment rate past 7 percent. Across the country, and particularly in the South, Black Americans escalated their long-running campaign of civil resistance to the regime of white supremacist violence that had always circumscribed their opportunities. Martin Luther King Jr. led the Southern Christian Leadership Conference in carefully selected local efforts to win power for Black Americans. The nation that elected John F. Kennedy was the richest and most powerful on Earth and possibly in the history of humanity, yet tremors moved through the vast structure.

For his secretary of the interior Kennedy selected Stewart Udall, an Arizonan of Mormon descent who had flown fifty combat missions as a B-24 gunner in World War II. Udall clashed immediately with Bureau of Reclamation commissioner Floyd Dominy, who nominally served under Udall but had thoroughly fortified his own position in the federal bureaucracy. Though a vicious tyrant and shameless philanderer, Dominy drew power from human relationships. From mailroom clerks to highly placed members of Congress, he had spent years memorizing names and delivering dam projects. He counted as his most valuable patron Carl Hayden, the extremely powerful senator from Arizona, about whom more later. Suffice it to say that Dominy mounted a photo of the pair wearing flower leis on a Honolulu tarmac on the wall above his desk.[2] Anyone who wanted to confront Dominy had to weigh his options under the leering grin of the chairman of the Senate Appropriations Committee.

Dominy was a builder at heart, a land baron drawing a government salary, but Udall saw the U.S. waterways as spigots that went to waste if left untapped. Udall, born in 1920, had come of age during the Great Depression and formed his political consciousness around Franklin Roosevelt's long reign. He believed in American greatness as expressed through bold acts that upended traditional hierarchies. While attending the University of Arizona, Udall and his brother, Morris, assisted Black students in a campaign to racially integrate the student cafeteria. After completing his education Stewart Udall opened a law practice, and, in 1954, he won election to the U.S. Congress, where his interests in conservation took a back seat to the realities of holding office in Arizona. Politicians who landed water projects got reelected; those who opposed them were held as oddballs and cranks.

Udall supported John F. Kennedy in his Democratic primary race against Lyndon Johnson based on the belief that Kennedy would be a transformational politician similar to Roosevelt and found himself rewarded when Kennedy won. Udall swiftly took advantage of his new position as secretary of the interior to force the racial integration of the National Football League (NFL) Washington DC franchise (then playing under the slur "Redskins") in 1961. The Department of the Interior owned the stadium where the football team played home games. Udall successfully threatened team owner George Preston Marshall into hiring Black players, finally joining the rest of the NFL as the last franchise to racially integrate.

Udall directed the National Park Service to recruit workers and rangers from historically Black colleges and universities. He reorganized the Bureau of Indian Affairs to serve the material interests of Indigenous Americans rather than acting as yet another agent of their oppression. He opposed Cold War military adventurism and toured the Soviet Union alongside beloved poet Robert Frost. Udall's deeply held commitments to equality, democracy, and American workers bring to mind his spiritual predecessor, Harold Ickes. Though their personalities were very different, Udall grew up in the country Ickes (and his contemporaries) created and both approached public service as a nearly sacred charge.

Udall rarely shied from conflict, but he knew he couldn't simply oust Dominy. The two men disagreed vehemently about the conservation movement and its role for the future. The Bureau of Reclamation commissioner knew that money ruled Congress, and the environmentalists didn't have much money. Udall and Kennedy had an investment in the nation's moods that no district-bound congressman could match, and they appreciated the clout wielded by the swelling throngs of environmentalists. Generations of intense industrial development largely unregulated by the government (first ambivalent during the Gilded Age and later desperate to beat the Depression and the Nazis) had deposited a truly sickening volume of toxic waste across the United States. Factories and communities dumped anything they liked into any convenient waterway and belched dangerous chemicals into the air, often in close proximity to housing.

This was not merely a big city problem: many suburban communities had been built as company towns around extremely dirty industrial plants. In October 1948 a cloud of toxic smog emitted by the zinc and steel mills of Donora, Pennsylvania, settled over the idyllic town. The smog suffocated Donora's inhabitants, burning the mucous membranes of their eyes, noses, throats, and lungs. "We still thought that was just normal for Donora," recalled a resident, since the pollution often got intense.[3] But whereas wind typically dispersed the smog, weather patterns now kept it anchored in place. For five days it darkened the sun, choking the young, the old, and the sick. By the end of the week at least twenty people were dead in the nation's worst ever air pollution disaster. The Cuyahoga River, which runs through the industrial sectors of Cleveland, Ohio, was so revoltingly polluted that the river itself famously caught fire on many occasions. The Cuyahoga wasn't even an exceptional case, though a 1970 *National Geographic* cover story ("Our Ecological Crisis") exposed it to nationwide infamy. It is hard for younger people like myself to imagine just how disgusting and dangerous living conditions had become in many American communities by the 1960s.

People noticed this. They observed decline in their lived environments, noticed the dead birds in their yards, the watering holes signed unsafe for swimming, and could not help but perceive the depleted stars of the night sky. In 1962 Rachel Carson published *Silent Spring*, a wildly popular account of the damage done to bird populations by agricultural pesticides. All across the United States, millions of people saw their environments under attack and responded to defend them. They had no shared ideology to unite them, or even a common cause exactly—most of these people had not ever been seriously involved in politics. Between Franklin Roosevelt's New Dealers and the remnants of Teddy Roosevelt's Republicanism, environmentalists identified with no specific party. They couldn't amass any real power in most Congressional districts, but they numbered in growing millions, and their nationally distributed votes were valuable to President Kennedy. At the same time, Kennedy and his inner circle came into the White House with few ideas for the West: he promised to approve new public power developments but had no specific plans and left Stewart Udall to formulate them.[4] Udall was a conservationist, and he expected his underling Floyd Dominy to get with the program.

To Dominy's understanding, he had been placed in office to expand and enhance the property portfolio of the United States. By that standard he was easily the best Commissioner in history, dedicating a spectacular new high-rise office building in Denver to support the bureau's swollen ranks of engineers, surveyors, and drafts people. An entire generation of engineers looked up to the Bureau of Reclamation and Corps of Engineers as the dynamos of American development, edifices of public service that also maintained a sort of cowboy panache. Why would Floyd Dominy back down to a transitory cabinet secretary like Udall? He saw no reason to change a damn thing. The United States was in the middle of a multigenerational campaign to green and expand the West—a project Udall understood well enough to ardently support the Central Arizona Project (CAP), an open-air aqueduct running several hundred miles over a mountain range and through a hundred-degree moonscape to deliver water from the Colorado River to the distant southern cities of Phoenix and Tucson. When it came to brass tacks for his own home state, the ordinarily self-righteous Udall reverted to his old Congressional habits. This, in Dominy's eyes, rendered Udall's environmental concerns little more than soft-headed bleating.

They Tried to Flood the Grand Canyon

U.S. senator Carl Hayden of Arizona served as a role model for a generation of U.S. senators to follow. First elected to the U.S. Congress in 1912 and the Senate in 1927, the "Silent Senator" almost never spoke on the chamber floor, correctly recognizing debate in the U.S. Senate to be entirely theatrical. Instead he focused on the distribution of resources: money, land, food, and water. The New Deal's diverse coalition operated by distributing financial favors to legislators' districts. Every city in America has a long list of projects they want to build but cannot afford to on their own. For much of the twentieth century, most representatives and senators measured their worth by how many of these little priorities they could deliver. Legislators like Carl Hayden or his House Appropriations Committee counterpart Clarence Cannon of Missouri commanded immense power without even the pretense of discourse. "No man," a fellow Senator quipped anonymously of Hayden to the *New York Times*, "has wielded more power with less oratory."[5]

Cannon was a colorful character himself. Like Hayden he

served into advanced age, wielding vast power even as infirmity forced him to cede ever more work to his staffers. Of Cannon, the *Times* noted archly, "he is not an inspiring orator. When words fail him, he is not above lashing out with fists. He fought twice with fellow House members and once with the late Senator Kenneth McKeller of Tennessee. The fight with Mr. McKeller . . . wound up with the two elderly statesmen chasing each other around the table with gavels as weapons. Nobody was hurt."[6] One can hardly imagine a scene more prophetic for the future of the U.S. Congress than two of the most powerful men in America chasing each other with pantomime weapons, each of them too decrepit to land a blow.

In the 1960s the state of Arizona wielded political power vastly disproportionate to its size. Despite containing less than 1 percent of the national population (1.5 million against 192 million), Arizona gave the nation its Senate Appropriations Committee chairman in Carl Hayden, its Secretary of the Interior in Stewart Udall, and a Republican presidential nominee in Barry Goldwater (who served alongside Hayden in the Senate).[7] Goldwater came to be seen as the avatar of a newly resurgent right wing in the Republican party, but his life had been shaped by government largesse. The senator was heir to a chain of Arizona department stores founded by his grandfather, who had immigrated to the state to sell clothes and equipment to the men working federal water projects. Decades later Barry Goldwater made his name in politics by vocally opposing the New Deal and organized labor. He hated the idea of the workers who had made him rich from birth organizing for better pay and conditions, which endeared him to Herbert Hoover and the deeply conservative Republicans of the West. He rose to the U.S. Senate, at which point he found his ambitions stifled by the fact that Senate Republicans were a minority caucus, and Arizonans knew Carl Hayden commanded far more resources than did Goldwater.[8]

Lacking any incentive to invest himself in governing, Goldwater began acting out: supporting Joseph McCarthy's red scare, warning of international Communism and domestic socialism, suggesting to bloodthirsty audiences that nuclear weapons ought to be used in theaters like Korea and Vietnam. Goldwater evinced an admirable personal commitment to racial equality (he integrated his family business when he took it over and integrated the Senate cafeteria on behalf of Black staffers) but nonetheless voted against the 1964 Civil Rights Act. Republicans rewarded him with the presidential nomination. Martin Luther King said of Goldwater in 1964, "I feel that the prospect of Senator Goldwater being president of the United States so threatens the health, morality, and survival of our nation that I can not in good conscience fail to take a stand against what he represents."[9] Goldwater left the Senate to run for president and lost handily to Lyndon Johnson in the 1964 election but remains an inspirational figure for Republicans to this day.

If Carl Hayden was one of the primary reasons Barry Goldwater lost interest in governing, he was also the major driving force behind the lengthy campaign to flood the Grand Canyon. That was not the goal, of course, just an inevitable outcome of building a massive dam at its base. Bridge Canyon Dam, the Bureau of Reclamation called it, drawn up 740 feet over the Colorado River's muddy bed. Bridge Canyon was (and still is) considered one of the very best dam sites remaining in the United States. The canyon itself is very remote, bare rock and desolation, all but inaccessible to rafters or hikers. Lined by walls of impermeable stone, the river takes a nosedive through a thousand feet of brutal cataracts—this steep drop through a narrow canyon makes for a relatively small dam that would enjoy at least 650 feet of hydraulic head. The ninety-mile-long reservoir planned behind the dam would serve no irrigation purpose, but it would allow the dam to act as a cash register generating electricity for sale on an eager market. There was just one problem: Bridge Canyon sits just downriver from

Grand Canyon National Park, and while the dam itself would lie outside of the park, that ninety-mile reservoir would inevitably back up into it. At the same time, a second dam would go up at Marble Gorge Canyon, just upstream of the Grand Canyon. The Glen Canyon Dam had already begun to impact the Grand Canyon's ecology by desilting its water (see chapter 5), but this plan would effectively lock one of the planet's greatest natural wonders inside a concrete cell. A portion of the lower Grand Canyon would be flooded indefinitely, converting the free-flowing river environment into a stagnant pool. A small price to pay, Carl Hayden thought, for the opportunity to finance the CAP's massive aqueduct to Phoenix and Tucson.

Floyd Dominy had a great relationship with Hayden. Both saw Bridge Canyon and Marble Gorge as crucial investments for the future: two of the finest, most productive cash register dams to be had anywhere, generating millions and millions of dollars that could be shifted anywhere Dominy desired within their state of origin. With those dams up and running, just about any project the bureau dreamed up could be made to look economically viable—Dominy would simply count the losses against the profits of the dams and expect Congress to nod along. What did they care? Sooner or later their district would need a multimillion-dollar water project of its own. The Grand Canyon dams, Hayden and Dominy thought, might even finance a project to capture part of the Columbia River and turn it south to California. Stewart Udall assiduously kept his name out of the authorizing documents but did nothing to halt Dominy's plans.[10]

The conservation movement felt that the Department of the Interior had betrayed them enough. Under David Brower's leadership the Sierra Club had attained a national profile and great credibility along with tens of thousands of new members, and across the nation few notions got regular people more indignant than flooding one of the nation's premier landmarks. The interstate highway system and gradually rising standards of living had put visiting the Grand Canyon in reach for many families, and they had taken advantage: from 1950 to 1966, the park's visitors tripled, from 600,000 to 1.8 million people.[11] Even those who'd never visited tended not to like the idea of flooding it. Floyd Dominy's accounting meant nothing to them, but neither did their objections matter a bit to Dominy. He had adopted a bunker mentality that let him disregard any inconvenient information.

Reader's Digest denounced the Bridge Canyon and Marble Gorge dams in 1966, demonstrating how mainstream America had turned against the idea. *Life* chased their competitors with a scathing article of their own, followed by children's classroom magazine *My Weekly Reader*. "You're in deep shit when you catch it from them," Bureau official Dan Dreyfus recounted to Marc Reisner. "Mailbags were coming in by the hundreds stuffed with letters from schoolkids. I kept trying to tell Dominy we were in trouble, but he didn't seem to give a damn."[12] Dominy could not make the numbers line up without his cash register dams, and he insisted on balancing his books even as others in the bureau advised him to simply run a deficit and get bailed out by Congress later. "Even though Dominy was a good liar when he had to be, here he was a prisoner of his own intellectual honesty," Dreyfus concluded.[13]

When the Colorado River Basin Project Act finally came in from the Congressional drafters in early 1968, the Grand Canyon dams were not in it. Stewart Udall understood the fight was lost, but the problem was selling it to Carl Hayden and Floyd Dominy. Hayden spent much of his time in the hospital by 1968 and announced in May that he would leave office at the year's end—he was eager to see the CAP receive its final approval and wouldn't balk at a creative solution. Dominy was a harder nut to crack, but, fortuitously, he was forced to spend that summer on a legally mandated whirlwind tour of the bureau's far-flung international facilities (they had begun exporting the American dambuilding model to the United States' client states abroad).

While Dominy was away, his subordinates nervously drew up a plan to authorize the CAP without the Grand Canyon dams. To provide the electricity needed to pump millions of gallons from the Colorado River over a mountain range to Arizona's southern cities, the bureau would simply buy a share of a traditional coal-fired power plant. Phoenix would get its water, and the Grand Canyon would stay pristine, but the product would not be self-financing. The bureau's dreams for Bridge Canyon and Marble Gorge were dashed. When Dominy returned from his trip, he signed the authorization his staff presented and ordered them out of his office without the slightest acknowledgment that he had been defeated. Carl Hayden gave the arrangement his blessing, declaring the CAP to be the most significant achievement of his long career. Stewart Udall must have thanked his lucky stars.

Given enough time in office, Udall might have been able to storm Dominy's entrenched position atop Reclamation. After all, Udall enjoyed such a close relationship with John F. Kennedy that the president dispatched his Interior secretary to conduct sensitive diplomacy with the Soviet Union. But then Kennedy died, and power shifted quickly within a Democratic Party that would never quite be the same. Though they helmed the most popular and successful political movement in American history, Roosevelt's heirs would begin to operate from a position of dour pessimism. Kennedy had been punished, they suspected, even if they weren't quite sure what for or by whom.

"The men are as different as day and night," Colorado congressman Wayne Aspinall fretted to Stewart Udall of the assassinated president and his successor, Vice President Lyndon Baines Johnson. "Kennedy [was] an idealist, and Johnson is a tough, brutal politician."[14] Johnson had earned his reputation over decades in Congress and the Senate, where he was ruthless in pursuit of his goals. The former Senate majority leader might bully or even physically intimidate during negotiations, but he was at heart a Texas horse trader: anything and everything became a bargaining chip. Ascending with the power of the New Deal Democrats, Johnson understood better than anyone alive how the resources of the U.S. government could be deployed to aid or harm a Congressional district. That made a cunning Capitol Hill operator like Floyd Dominy valuable to Johnson, more so than the idealist Kennedy. Udall would stay on at Interior under President Johnson, but any proximate threat to Dominy's position died with President Kennedy. The two men simply had to live with each other until 1969, when President Nixon took office and replaced them both.

Tocks Island

Through the backyard of the Maryland home where my father grew up runs a natural creek, a beautiful little bit of nature and a safe corridor through the neighborhood for dozens of native species. Yet for decades it was clad in fractured concrete and used as a duct for floodwater that could rush unpredictably down the skate ramp built for it—a perfect little thing turned snarling and ugly. Engineers now appreciate that natural waterways and green belts control floods by slowing and trapping the water. Waterways are larger than they appear, as water flows underground as well as over the surface. The "slow water" nourishes the land, plants, and animals while protecting people from dangerously hard-rushing culverts in their backyards. Over several years, crews from the state of Maryland ripped up the old concrete behind my grandparents' house and engineered in its place a beautiful, natural stream that manages water instead of blasting it as quickly as possible from one place to another.

That old one-story house stands within the borders of Baltimore County, so it draws its tap water from three surface reservoirs, each backed up behind a concrete dam. There are no natural lakes in the entire state of Maryland—all of them are manmade.[15] Baltimore County pulls from the Liberty Dam on the Patapsco River (built 1954) and two dams on the Gunpowder River: the Loch Raven Dam (built 1915) and the Prettyboy Dam (built 1932). In times of drought the county may also pump in water from the Susquehanna River to the east. The Susquehanna is one of the great rivers of the American East, deep and wide and lined by towering trees. The woods growing along its flanks are deep, dark, and quiet. Bald eagles traffic the sky over the gorge the river has made, and the I-95 interstate highway hops the river on a magnificent suspension bridge high enough for communion with the birds.

That highway is named for John F. Kennedy, one of many edifices named for the country's fallen young president, and, if you follow it eastward, you will eventually reach an even greater river—the Delaware, a leviathan of commerce and industry that snakes through the most densely populated region of the United States. George Washington led the Continental Army across the ice-choked Delaware for a daring winter sortie against Hessian mercenaries in the city of Trenton. One in ten Americans draws their drinking water from the Delaware and, without it, the metropolis of Philadelphia would not exist. It is also, despite the best efforts of Stewart Udall, Floyd Dominy, and the federal government, the largest river east of the Mississippi never to be dammed.

As the 1960s progressed, the environmental movement began to successfully obstruct the government's plans where before they had proved only minor obstacles. Previous generations of dambuilders had the advantage of flooding largely empty and remote landscapes like Lake Mead behind the Hoover Dam. Where they met opposition it was disorganized and lacked the resources to rebut the government: the bureau told people what a dam would deliver, and they had to take the engineers' word for it. Many reservoirs destroyed land belonging to Indigenous Americans, but nobody thought twice about double-crossing them.

Twenty years after the end of World War II, citizens were much more sophisticated. Many had gone to college and knew from their union jobs how to organize others. Movements that had once been so powerless that they could only beg the Bureau of Reclamation to only flood Glen Canyon and spare Dinosaur National Monument could now refute the government's claims on their own terms. At the Red River Gorge in Kentucky, on the Meramec River in Missouri, and elsewhere across the United States, environmental groups found they could effectively halt dam projects. Even if Congress approved funding, time and inertia was on the conservationists' side.

Like most American rivers, the Delaware had flooded in the eighteenth and nineteenth centuries, occasionally to destructive effect. The Army Corps of Engineers made plans to dam the river near its high tributaries in the Pocono Mountains along the Pennsylvania–New Jersey border in the 1920s and 1930s, but the notion had little momentum until August 1955, when two hurricanes hit the mid-Atlantic coast in the space of a week. The first, Hurricane Connie, ruined crops and killed nearly 30 people in North Carolina on August 12 before depositing a huge amount of rain in a northwestern swath reaching up to Ontario. Just five days later, Hurricane Diane came inland at nearly the same point. She took an unusual course inland, looping north and east and back out to sea near Cape Cod, and her rainfall spurred the already rushing waterways of the Delaware River basin to a full stampede. Earthen embankment dams in northern Pennsylvania and New Jersey overtopped or ruptured entirely, releasing torrents of floodwater that destroyed more dams and even entire towns. Camping sites packed with late-summer vacationers became horrific scenes as flash floods swept away tents and buildings with families still inside. Waterways from the Carolinas to New England leapt from their banks to savage adjacent cities. Officially Diane killed 184 people, left thousands hospitalized, and tens of thousands more homeless. At the time, Diane was the most costly hurricane disaster in American history: the first storm to clear $1 billion in damages.

Naturally the federal government stepped in to try and prevent a repeat disaster. In the years that followed, Congress encouraged the Corps of Engineers to find a solution. The corps thought big and aimed high, proposing a network of dams on the Delaware's tributaries anchored by the largest dam east of the Mississippi. Just 160 feet high but 3,200 feet long, it would choke the river at a geographically advantageous spot called Tocks Island. Lushly forested cliffs rise hundreds of feet above the river gorge. The Tocks Island reservoir would inundate twelve thousand acres of river gorges and farmland to a depth of 140 feet, backing up water for thirty-seven miles upstream.

The corps envisioned an inland sea, at once protecting the Delaware River basin from disaster while securing drinking water for residents of four states and establishing a 70,000-acre national recreation area with hundreds of miles of new shoreline around the reservoir. Ambitiously, they anticipated ten million Americans would visit every year. Legally speaking the agreement had no precedent: the federal government had moderated agreements between the states before, as with the Colorado River Compact of 1922, but never had it entered into a compact as an equal partner with its own constituent states.[16] Congress approved initial funding for the Tocks Island Dam in 1962 as part of the Flood Control Act, and President Johnson signed the legislation establishing the Delaware Water Gap National Recreation Area on September 1, 1965.

A decade earlier, that bill might have ended the discussion. "Had the Corps moved less ponderously," the *New York Times* suggested in 1975, "a start on construction may have decided the question a decade ago before environmentalists formed a coalition and geared up for the Tocks fight."[17] But the corps could not start work in 1965 due to another Johnson administration priority: the war in Vietnam. Congress had passed the Gulf of Tonkin Resolution the year before, authorizing a major escalation of the war, and the Army had more pressing uses for nearly $100 million than a mid-Atlantic water project.

In the meantime, the corps started acquiring the land to be submerged: more than six thousand plots housing four thousand families, some of whom had been working their land since the 1700s. The politics were awful, but the corps did not answer to voters. They forced those families to sell prime agricultural bottomland at shameful prices and sent federal marshals with shotguns to clear out those who objected. Corps of Engineers workers quartered themselves in some of the empty homes (they did pay rent to the owners), which rubbed salt in fresh wounds. In the following months and years, a small number of people from outside the community took up unauthorized residence in some of the empty homes and began tilling the fallow fields. When the corps tried to force them out by demolishing the structures they were using, they positioned themselves and their children as human shields.

Wary of further bad press, the corps sought a court order for forcible eviction, but this only made them look worse to the locals as they seemed to be granting more consideration to the "squatters" (an abhorrent word to describe anyone, but it is ubiquitous in contemporaneous news coverage) than to the longtime property owners the corps had just evicted. The government finally resolved the issue in 1974 by sweeping up all unauthorized residents on a freezing cold night and depositing them on a remote highway shoulder while bulldozers and marshals destroyed everything they owned. This made the corps look cruel even to those locals who resented their illicit neighbors. They could not possibly have handled everything worse.

Leading the local opposition were two Pennsylvania housewives named Nancy Shukaitis and Ruth Jones. Shukaitis, who passed away in March 2021 at the age of ninety-six, recalled a 1964 hearing where a local mayor led his audience astray: "He showed slides of bodies covered with sheets. 'These were the people who died in the 1955 flood,' [we were told]. When I returned home, I wrote to all community officials between Port Jervis and Trenton and my opinion was confirmed. There were no lives lost on the mainstream of the Delaware—only on the inland tributaries."[18] That being the case, the Tocks Island Dam (had it existed in 1955) would not have saved anyone, and building it would leave genuine dangers on the tributaries unaddressed.

Ruth Jones found her political courage facing down a U.S. marshal who came to convince her family to sell their home: "We stood there in silence . . . my daughter, who was the youngest, stood there and the tears were rolling down her cheeks. . . . I looked at [her] and I got *so angry*," Jones recounted to a camera in 2006. "I got my German temper up, I got those papers and I shoved them back in the hands of that U.S. Marshal. I says, 'tell [the Corps of Engineers] to shove these up their ass!'"[19] In the end they were forced out, but they maintained several canoe rental shops on the Delaware and continued to fight the project. Jones and Shukaitis formed the Delaware Valley Conservation Association, organized their neighbors, and enlisted allies with national platforms. Supreme Court justice William Douglas traveled to the Poconos for a well-publicized pilgrimage to Sunfish Pond, a place of serene beauty that the dam would have erased from the Earth.

Even people who did not count themselves conservationists jumped into the fight, convinced the Tocks Island Dam would cause far more problems than the Corps of Engineers was willing to admit. To start, many suspected that the region's bedrock could not support a dam of such size. The existence of the Delaware Water Gap—where the river has punched a hole clean through a mountain—suggests on its own that the local rock cannot stand up to water, and a study from 1939 concluded that a dam at the site would likely fail due to geology. On top of that, seasonal releases of water from the dam risked exposing acres of inundated land to open air, rotting trees and all. These areas would quickly become disgusting messes whose half-decomposed runoff would inevitably pollute the reservoir once it refilled. Compounding the pollution issues, farms and chicken processing plants from upstate New York dumped their refuse directly into upper Delaware tributaries. Free-flowing water diluted these pollutants enough to keep the river healthy, but a stagnant reservoir would let them accumulate. The more anyone read or heard about the dam, the worse it seemed.

Meanwhile, national politics had grown very dark. President Kennedy's assassination inaugurated not just the Johnson administration but a horrifying era of political violence. Martin Luther King, Malcolm X, and Robert Kennedy were the highest-profile figures to fall, but across the country the FBI and local police forces were engaged in active campaigns of terror and murder against civil rights activists, communists, or anyone suspected of consorting with either. The Civil Rights Act of 1964 granted full political rights to millions of people for the very first time in American history, but the white response in many places (especially the South) was vigilante killings and church burnings. Medicare and Medicaid, two of the most successful public health programs of the twentieth century, came into being in 1965. Johnson's government was spending more money than any administration since World War II, much of it on programs that were supposed to make America a kinder and safer place. Yet to many people, even New Deal Democrats who supported Johnson's priorities, everything seemed to be getting worse.

The war in Vietnam conscripted hundreds of thousands of young Americans into a horrific conflict, and Johnson felt he could do little except lie to the public about how badly South Vietnam (and by extension the United States) was losing the war. In 1968 Johnson dropped out of the Democratic primary race, advancing Vice President Hubert Humphrey to the nomination as Roosevelt's party tore itself to pieces over Vietnam. How could the party of the Civil Rights Act also wage an unjust and immoral war abroad, spilling Vietnamese blood using countless Black, Latino, Asian, and Native American soldiers conscripted against their will? It was untenable, and Humphrey faced a ruthless opponent in the general election. Former vice president Richard Nixon ran a famously dishonest campaign against Johnson, not his actual opponent in Humphrey, and won the election handily.

Nixon's total lack of principle occasionally led him to support a decent outcome. He'd given environmental issues almost no consideration whatsoever in his campaign, yet an oil spill in his native California occasioned Nixon to declare his "concern for preserving the beauty and the natural resources that are so important to any kind of society. . . . We are going to do a better job than we have in the past."[20] Whether he truly believed this or, rather, saw a low-risk avenue to a few more votes, the new president actually followed through for the environmentalists. Congress established the Council on Environmental Quality (CEQ) in January 1970, but the Council was placed under control of the White House, and it could not have performed its role had Nixon not empowered the agency.

It was the CEQ that put the screws to the Corps of Engineers over Tocks Island. At first the corps refused to acknowledge environmentalists' concerns, but the National Environmental Quality Act (signed by Nixon in 1969) mandated environmental impact statements for all major federal projects. After some conspicuous prodding by the CEQ, the corps finally admitted in 1971 that the Delaware River environmentalists' concerns were valid. The council recommended that Congress freeze funding to the Tocks Island Dam unless answers to those concerns could be found.

The politicians all pulled in different directions. The governor of New York declined to spend the money necessary to control agricultural runoff from his state, and the governor of New Jersey demanded the scale of recreational amenities be slashed to limit visitors to four million per year, so as not to saddle small communities with outsized costs to upgrade their infrastructure. Meanwhile, the governor of Pennsylvania declared the Tocks Island Dam to be necessary for the state's future. The Department of the Interior avoided weighing in too vocally lest others accuse it of throwing its weight around, but, in 1972, it acknowledged the recreational area could still operate as a park surrounding the free-flowing Delaware River.[21]

By 1975 the bill authorizing the Tocks Island Dam was approaching its ten-year sunset. Congress had to decide what to do with it, so they asked the project's principals for a vote. Of the four states that had initially signed on, three governors (New York, New Jersey, and Delaware) now opposed it, with only Pennsylvania voting to proceed, and the Department of the Interior abstaining. The dam was dead. Congress handed over the lands already acquired to the National Park Service for conversion into the Delaware Water Gap National Recreation Area as it exists today: forests, meadows, lakes, trails, campgrounds, and the unencumbered Delaware River. The Corps of Engineers redirected their flood control efforts to smaller dams and reservoirs on the Delaware's upper tributaries. While the park that stands today is very nice, those locals who lost their lands remained embittered, and the corps never quite recovered from their sordid record in the Tocks Island Dam affair.

The dam was clearly a bad idea, yet, if not for a few accountability laws passed in the early 1970s, the public might have been suckered into this terrible proposition by a bunch of smooth-talking engineers who had not done their homework. How often, one might reasonably wonder, had they actually pulled it off under the public's nose? After Tocks Island, local groups were tougher, fiercer, and better prepared. They had a real blueprint to beat the feds at their own game. The benefits the bureau or the corps would credibly claim from big dams were falling, while the apparent costs only climbed skyward. As time went on, what had been concrete-clad investments increasingly seemed like boondoggles. One can't say precisely when the golden age of big dams concluded, but the foundering of the Tocks Island project in 1975 demonstrated its time had passed.

Nixon and the EPA

Richard Milhouse Nixon was a cynical liar and a terrible president on the whole. There can be no doubt whatsoever about that. The events leading to his resignation have been so thoroughly covered by so many authors that nobody needs to read them again here. Nixon's ruthlessness, for all its drawbacks, meant that when his administration set a goal they used every tool at their disposal to achieve it.

In 1971, responding to a rapid fall in the value of the American dollar, Nixon declared that dollars would no longer be pegged to the international price of gold, as Western industrialized nations were supposed to do. A dollar was now worth whatever the Federal Reserve Bank said it was. The White House also issued a series of executive orders that established price controls over gasoline and meat, which had grown painfully expensive for many Americans. The government was sharply divided over these moves, and economists have debated for decades whether these interventions were correct in the long term, but most newspapers and voters thought it a great success. Nixon understood that whatever the economists said, people wanted their government to act in defense of their material interests. From his perspective, if public opinion backed him up then he would never have to answer to anyone else.

The year before, Nixon had capitalized on the earlier success of his CEQ by following the council's recommendation to establish the Environmental Protection Agency (EPA). A cascade of environmental disasters in the 1960s had convinced many millions of Americans that the country needed to change its practices. Former CEQ and EPA official Chuck Elkins recalled, "You could say that the year 1970 was the heyday of the environmental consciousness in this country. . . . 20 million people participated in [the first Earth Day] teach-ins all over the country [out of 200 million]. So, one in ten people participated in the first Earth Day. . . . I have to say, it was somewhat of a mystery to me, even though I was living through it. You know, it felt as if those of us working in the environment were working in sort of a small area . . . [not of] interest to the American public. And then all of a sudden everybody is talking about the environment."[22]

Nixon was not about to be left behind on this trend. He issued the nation's first Presidential Environmental Message in February to set out his environmental agenda in thirty-seven bullet points: new national standards for air quality and automotive emissions, a tax on leaded gasoline, clean-ups of federal and military facilities that had polluted in the past, four billion dollars for water treatment facilities, and quite a bit more. The president sent Congress a plan to consolidate the scattered environmental functions of the government into the EPA. By the summer Nixon's proposal was law. He selected as the EPA's first administrator a man named William Ruckelshaus, an assistant attorney general who in Nixon's last days would resign as part of the infamous Saturday Night Massacre.

Whereas prior programs had focused on directly mitigating or cleaning up polluted federal lands, the EPA enjoyed a broader mandate. To this day the agency determines not only which lands are polluted but conducts its own research into pollution and has the power to determine which substances even count as harmful. It establishes environmental baselines in each locale, monitors deviations from those baselines, and provides training and resources to state governments to help maintain environmental standards. If properly empowered it performs an absolutely vital function for a modern industrialized society. Ruckelshaus believed fervently in his agency's mission and enjoyed President Nixon's trust. He was, by all accounts, a very capable and conscientious administrator—so respected that, in the 1980s, President Reagan would tap him to lead the EPA once again.

Again, the Watergate Complex break-in and subsequent

coverup orchestrated by President Nixon is a very complex story that has been extremely well documented, and we will not wade into the proverbial reeds here. Ruckelshaus became involved in April 1973, very late in the proceedings, when acting FBI director L. Patrick Gray resigned following the revelation that he had destroyed evidence from one of the convicted Watergate conspirators. White House officials were resigning left and right, and the administration was in tumult; amid the shuffling, Nixon tapped Ruckelshaus to temporarily run the FBI. The gig was a long way from the EPA, but Ruckelshaus had worked for the Department of Justice and had a sterling reputation for integrity. Six months later, Nixon felt increasingly cornered by special prosecutor Archibald Cox, who was closing in on audio tapes that would ultimately incriminate the president. Nixon ordered his Attorney General, Elliot Richardson, to remove Cox from his position. Richardson resigned rather than follow the order. In a rage, Nixon passed the order to Ruckelshaus, who refused it and submitted his resignation as well. Solicitor General Robert Bork, a right-wing operative lacking any scruples, eagerly stepped in to do Nixon's bidding.

Nixon had turned the Republican party into a criminal enterprise, welcoming an element that has never left the party even if it has not always enjoyed total control. All the decent men had departed by October 1973, though Nixon would not crawl away from the White House until August 1974. The following month, the hapless new president, Gerald Ford, issued a preemptive pardon for all Nixon's crimes. Ford was a paper man blinded by the personal affection he felt for the personally odious Nixon. "I had no hesitancy about granting the pardon," Ford later told journalist Bob Woodward. "I felt that we had this relationship and that I didn't want to see my real friend have the stigma."[23]

Immediately pardoning the criminal ex-president was the political equivalent of sprinting face-first into a garage door. The public was furious at Ford, along with the Democrats who still controlled both chambers of Congress. Ford would spend his stunted half-term presidency turning in circles, immobilized in the lacuna Watergate had created at the heart of American politics. Everyone was angry, justifiably so, and yet the prospect of retribution or even a semblance of justice had been ruled out of bounds by this unelected stooge (for Ford had not even shared the presidential ticket with Nixon in 1972 and could claim no legitimacy). Ford trailed badly in polling for the entire 1976 campaign, though in the end low turnout kept the results close. Looking at the state-by-state map, the geographical alignment is striking, arrayed entirely around the Mississippi River. All but a handful of states east of the Mississippi went for Carter, and all but two states west of the Mississippi (Texas and Hawaii) went for Ford.

American democracy, which had enjoyed three decades of unbroken ascendance, had watched the New Deal regime end only to be replaced by a kind of empty, exhausted cynicism. In 1975 the Church Committee hearings in Congress revealed a sordid decades-long history of international assassinations by the CIA, which to many observers resembled the string of political assassinations they had recently witnessed in the United States. From these killings to the rough handling of families at Tocks Island, a distinct pattern of government-directed brutality became apparent. Whether the American government simply murdered its foreign and domestic opponents became a question reasonable people asked each other. American citizens were increasingly unsure the government looked out for their best interests anymore. Reassuring them that the federal government could improve their lives was the monumental task that fell to President Jimmy Carter.

7

THE BREAKDOWN

Dams collapse the same way most big things do, the way relationships fall apart, or families end up destitute: slowly, and then all at once. They look so solid that it is hard to imagine anything short of an atomic bomb or an historic earthquake that might sunder them. In reality a dam is an object in constant motion, merely one component of a dynamic system that pinions vast masses against one another. The largest and most important force in this system is not the dam but the reservoir of water behind the dam, constantly pressing against it. Like an overmatched boxer reclining on the ropes, the dam contains this force by leaning on the land around it. In the end a dam is only as strong as the ground on which it is built, and, in 1976, the geology of southern Idaho combined with institutional hubris to cause the worst dam failure since William Mulholland's St. Francis Dam collapse fifty years earlier: the Teton Dam disaster.

The state of Idaho, gorgeous and rugged and generally ignored by the rest of the country, is a cross-section of the Rocky Mountains that has been rudely cut in half. Great peaks and forests stand in the north of the state, dominating its narrow panhandle, but, in the squared-off south, the land quickly becomes flat and dry to the point of desolation. The primary source of water in this high-altitude desert is the Snake River, which tracks an ancient lava flow in an east-west crescent across the state on its way to join the Columbia River in Washington. Eons of moving water and volcanic sediment have deposited along Idaho's Snake River Plain some of the best growing soil in North America. It is porous and permeable to water, draining fast in precisely the way that many crops enjoy—especially potatoes, for which the state is most known. Agriculture dominates Idaho's economy, and most of that agriculture drinks from the Snake River or its many tributaries.

The Snake River pales in comparison to the Colorado, but a lesser dragon is a dragon still. Precipitation and runoff vary greatly from year to year and especially within each year, leaving growing conditions unpredictable. Idaho's early potato farmers employed the backbreaking-yet-effective irrigation techniques handed down by the Mormons, but one good monsoon could send enough water to blow through their sluice gates and ruin their crops. The year 1961 saw a historic drought on the Snake River Plain and 1962 an historic flood. In the rapidly growing American West of the mid-twentieth century, that kind of insecurity was unacceptable, and Idaho's politicians demanded their due in the form of federally funded water projects.

The Bureau of Reclamation targeted the Teton River for damming in 1963. Springing from the gray stone of the Rockies near the Idaho-Wyoming border, the Teton hopscotches down a steep, winding canyon some sixty miles to join a larger tributary called Henry's Fork that empties into the Snake River. The Teton can surely flood in a bad storm, but the fact that the irrigation-focused Bureau of Reclamation took on the work instead of the flood-focused Army Corps of Engineers demonstrates that this was first and foremost an economic project. Never mind that the farmers of the Snake River Plain pulled in profits year after year, or that their potato yields had actually increased during the drought year, or that some of these growers were already irrigating their crops with over *ten feet* of water per year, the equivalent of a tropical rainforest's rainfall. If California's Imperial Valley and a dozen other agricultural centers across the West deserved cheap government-supplied water, so did the potato farmers of Idaho. At their behest the bureau made plans to dam the Teton River, though the Vietnam War would delay funding into the 1970s. By that date environmentalists had learned the government did not play fair when it came to development. They fought the bureau in newspapers and public meetings, insisting the Teton Dam would destroy a scenic mountain river only to lavish irrigation water on farmland that already consumed more than its need.

The activists lost. Growers dominated Idaho politics, and they would have their water subsidies. Those opposed insisted that the state's politicians and major media organs had to be outsiders and communists. The Bureau of Reclamation broke ground on the Teton Dam in 1972 and spent the next three years impounding the river behind a wall of dirt and impermeable clay. Dam opponents had voiced concerns about the geological conditions at the dam site, fretting over what might happen in one of the region's frequent earthquakes, but Idaho's Congressional delegation halted a scientific review and forced the project through.[1]

That intervention would prove a disastrous error, for the ground beneath the Teton Dam was rotten. Eons of earthquakes and volcanoes left it pockmarked with countless cavities, fissures, and faults. Much of the rock the bureau assumed to be solid had been formed from heaps of volcanic ash compressed over thousands of years into a porous stone called tuff. Silicon dioxide crystals left the rock sandy and glassy, easily pulverized and reordered when earthquakes rolled through. With a reservoir's worth of water behind the dam, some would inevitably work its way into those gaps. All reservoirs seep and leak to some degree, but allowing water to rush through open spaces rapidly erodes bigger holes. Typically a large dam is anchored at its base to a section of naked bedrock—stone sealed against stone—but the Teton Dam would rest atop a foundation of sand.

The U.S. Geological Survey prepared reports and memos

advising the Bureau of Reclamation on major faults near the proposed dam site, especially on the dam's right abutment. By the time these reports reached bureau leadership in early 1973, their cautious scientific rhetoric had been watered down, and work on the Teton Dam was already underway. A dangerous dam site now looked merely risky, and the bureau had a hundred engineering solutions to mitigate that risk. "Every dam site we've built on [since 1940] would probably have scared the hell out of a nineteenth-century engineer," former regional bureau director Pat Dugan told journalist Marc Reisner. "But you probably wouldn't feel safe going a hundred miles per hour in a Model T either . . . you might feel perfectly safe in a Porsche."[2] The bureau trusted technology to intervene where prudence failed.

At the Teton Dam site they leaned on a favorite technique: grouting. When geologists find faults near a dam, they drill a hole and attach a hose to the opening. Into the hole goes a mix of water and cement (grout), injected at firehose pressure for hours or even days on end—however much grout it takes until the hole fills, sealing every crack and fissure without the need for exhaustive surveys. Drilling and grouting in systematic fashion along the edge of a dam establishes a "grout curtain," a subterranean barrier through which water cannot move. The Teton Dam's engineers grouted faults as they discovered them and coordinated their efforts poorly, a piecemeal approach that belied overconfidence in the geology. They treated it like a good site with a few problems, not a bad site that would challenge any engineer on Earth.

In February 1974, with the groundwork nearly complete but the project lagging behind schedule, a work crew made a frightening discovery. They had been excavating a foundation trench, a tunnel underneath the dam that gets packed with impermeable material before the reservoir fills in order to keep water from slipping under the dam. In the course of their work the crew discovered what Bureau of Reclamation project engineer Robert Robison's report described as "unusually large" fissures in the right wall of the river canyon.[3] A civilian would have called them caves, or even caverns: some were over one hundred feet long and wide enough to admit a Ford pickup. Everywhere excavators looked on that right abutment, they found another hole. Yet Robison somehow convinced himself all was well, telling his superiors, "We do not recommend to grout these voids at this time . . . as they are located outside the dam area and could be grouted at a later date."[4]

The worst voids on the right abutment went unfilled, as they lay slightly beyond the threshold of where the bureau thought water could reach. Other cavities around the dam site consumed half a million cubic feet of grout, more than twice what the bureau planned before construction. Each hose could pump three hundred gallons of grout-infused water per second, yet day after day they emptied themselves into holes that simply refused to fill. October 1975 saw the dam nearly finished. The bureau closed its diversion tunnels, arresting the Teton River's lifecycle for the first time. Robison had only to wait for spring's melt. A wet stormy winter promised enough runoff to satisfy the greediest irrigators.

By March 1976 river water rose quick as a drawn bath against the 305-foot dam. Reclamation safety protocols called for reservoirs to be filled no faster than one foot per day, but given the rapid inflows on the Teton River, Robison's request was approved to fill at two feet per day. Robison even argued the doubled fill rate would more rigorously test the bureau's grouting procedures. In a way, he was right. Engineers who drilled test wells to track any changes in the local water table found a rate of upwelling *one thousand times* the expected rate, but both they and Bureau leadership in Denver agreed that the rate was so high that the reading could not be trusted. It must be a "pressure response," they decided, a phenomenon akin to boosting the pressure of a garden hose stream by pressing your thumb over the opening.[5] Every warning light blinked red, yet the bureau remained utterly confident. Even had they recognized the looming catastrophe, they couldn't have done much to avert it, for the dam's emergency spillway still stood unfinished. The sink filled relentlessly with no stopper to pull.

The date June 5, 1976, still stands as the bureau's day of infamy. The reservoir lapped as high as it would ever rise. Robison and his team were anxious, having discovered over the prior days several consistent leaks at the base of the dam's right abutment. A leak in a dam isn't necessarily cause for alarm but something to keep an eye on. If the water was clear, it meant the leak ran clean through the ground without picking up sediment or scouring new pathways; dirty water meant the dam's innards were finding their way to the outside. Robison had been glad to see these leaks were clear the night before, but on the morning of June 5 a stream of filthy water spurted from the right abutment. A bulldozer attempted to stop the hole with stony rubble (a substance called "riprap") only to watch it grow with each passing minute. More leaks broke loose, closer and closer to the edge of the dam, until an unmistakable wet spot began to form three-quarters of the way up the dam's face. More bulldozers rushed to the spot, rolling at steep angles down the dam toward what quickly became a gushing sinkhole. All of this was captured by Bureau cameras; you can watch it in YouTube today. The sinkhole ate the rubble the dozers dropped and spat it down the dam. At this point Robison got the local county sheriffs on the phone and informed them they needed to evacuate the five towns located downriver.

It was 11:15 in the morning on a Saturday, and thank God for that. Had the Teton Dam failed at night or on a bustling weekday with parents at work and children at school, getting twelve thousand people to safety would have been nearly impossible. As it stood, the residents of the town of Wilfork, Idaho, had barely an hour. Given the timing most of them were already awake and going about the weekend. They knew where their families were and could quickly reach their vehicles once the alert went out on radio and television.

Upstream at the dam site, the bureau workers were evacuating themselves. Mud erupted from the sinkhole; the Teton Dam was in its death throes, vomiting out its own guts. The bulldozer drivers fought the flood with astonishing courage, pushing their machines well after they were clearly doomed, staring down the leviathan until the last possible moment. As the sinkhole sucked the dozers down, these men leapt from their seats, sliding on their rears down the dam face until their colleagues hauled them to safety by their rope harnesses. At 11:57 a.m., the entire right side of the Teton Dam collapsed with the force of a bomb. The entire reservoir tried to escape at once. To place this event in context, consider that Niagara Falls typically discharges around 170,000 cubic feet of water per minute.[6] When the Teton Dam collapsed, it disgorged 2,000,000 cubic feet of water *per second*.[7] This was one of the most explosive floods in all of human history. Water rushed out so fast it didn't even foam—it *stretched* like melted plastic into the space beyond until the canyon walls kicked it into a 150-foot maelstrom of violence.

The flood traversed the 6 miles between the Teton Dam and Wilfork in twenty-five minutes. Evacuees drove to high ground up a long road with views of the river, and they described the flood much like the people of Johnstown, Pennsylvania, had over eighty years before: less a body of water than a solid mass, a rolling hill of dirt and trees and other debris. It swept Wilfork away and scoured the town to bedrock before flooding Sugar City and Salem. Rexburg was farther away but stood immediately downstream of a large lumberyard, the contents of which became waterborne weapons that destroyed 75 percent of the town's structures. Thousands of acres of fertile farmland—the same fields the bureau had tried to subsidize by building the dam—were stripped of soil. Three hundred square miles of land flooded as the wave Teton Dam released traveled downriver for 155 miles. Only the American Falls Dam, itself known by the bureau to be unsafe under flood conditions, averted more destruction by dumping its own reservoir as fast as possible to absorb the fresh influx.[8]

That the breach had occurred in a sparsely populated corner of Idaho would limit its cultural impact. My mother, attending university in neighboring Washington state in 1975, could barely recall the event when prompted. Eleven people died. The damage to the Teton River gorge and to the Bureau of Reclamation's reputation would prove permanent. "It's kind of irrelevant," an Idaho Falls resident said in 1977 of the bureau's pending report on the dam site's safety. The same man described Teton Dam as "a classic example of a pork barrel water project . . . it was a bad one. We've built most of the good dams, and there's no compelling reason for more."[9]

Backpedaling

By the late 1970s the American psyche had been thoroughly brutalized. A decade of scandals and lies from the highest levels of government had undermined every institution in the land. The Central Intelligence Agency admitted to murdering opponents abroad, the Federal Bureau of Investigation was credibly accused of doing the same domestically, and one could not easily condemn the authorities as violent crime climbed to an all-time high in many cities. Presidents Kennedy, Johnson, and Nixon had all spent their tenures lying about the war in Vietnam: its extent, its progress, and their motivations for continuing it. Since President Eisenhower retired, the nation's leaders had been assassinated, deposed by their own party, resigned in disgrace, and lost their election following a pardon of the prior president's criminality.

The United States' position atop the world felt less secure. In 1950 the United States represented 40 percent of the recorded world economy. By 1977 Europe had rebuilt from World War II, and nations around the world from Brazil to China had rapidly industrialized, shrinking the U.S's share to 10 percent. Not a failure on America's part, but it was still experienced by Americans (and especially by American corporations) as a serious decline relative to the rest of the world. Inflation north of 5 percent enraged consumers and financial markets alike. Corporate profits declined by a third from their 1960s peak. The rate of profit to capital investment fell from 9 percent in 1966 to 4 percent in 1977.[10] The United States enjoyed the highest standard of living in the world but no longer enjoyed total domination over the global economy. At a certain point the elites making U.S. economic policy grasped this disconnect, though they carefully obscured it behind specialist jargon. The American people, they began to believe, were living beyond their means and needed a haircut. How exactly to sell that haircut in a democracy was an important question to which well-funded right wing organizations now dedicated themselves.

Jimmy Carter trusted himself to fix all this. There was no one on Earth he trusted more. Carter grew up a serious adherent to the Baptist tradition, and today his religious faith is probably his best-known trait. He studied engineering at the U.S. Naval Academy and served as a command officer and nuclear engineer on the Navy's new fleet of nuclear submarines before leaving the Navy to work his deceased father's Georgia peanut farm. Carter joined the analytical approach of an engineer to the evangelical certainty of a Baptist pastor. He intently studied important issues in order to arrive at conclusions he felt were backed by empiricism and science, at which point he expected everyone around him to fall in line and agree.

Though he spent a brace of two-year terms in the Georgia State Senate at the start of his political career, Carter found his true political identity as the state's governor between 1971 and 1975. He spurned the backroom dealing and horse trading that drove the legislature, preferring to be seen as standing above the fray. He saw personal virtue and morality as his biggest advantages, banking that he could convince the electorate to trust him more than the rest of the government and thus trust his policy proposals. Carter demanded more authority than previous governors had enjoyed to unilaterally overhaul Georgia's various bureaucracies, collapsing dozens of offices into each other in the name of budget savings that never materialized (historian Roger Freeman asserts in his 1982 book, *The Wayward Welfare State,* that Carter did little more than rearrange the proverbial furniture).[11]

Of course the Georgia legislature resented Carter's attempts to invest all the executive branch's functions in his person, as he refused to acknowledge their priorities. The Army Corps of Engineers planned a dam on the Flint River to mitigate the risk of flooding and provide an economic boost to southwestern Georgia, but Carter defied the state legislature by vetoing the bill. All the planning and politicking that had gone into the project meant nothing to Carter, who had personally studied the project and determined that the corps was underselling the dam's costs while overselling its benefits. He was probably right (and returned to the fight in 2008 to oppose a fresh round of Flint River dam proposals), but politics isn't really about being right.[12] Much of the Georgia legislature, to say nothing of the state's national Congressional delegation, was furious with Carter. Governors did not typically veto water projects anywhere in the country, but Carter stood firm, and today the Flint River runs gorgeously, lushly undammed, into the Apalachicola River and from there to the Gulf of Mexico.

Thankfully for Carter's Georgia opponents, the state constitution limited him to a single term, and, by 1975, the Baptist with the megawatt grin was a free agent with his eye on national office. His 1976 presidential run was thought a long shot until suddenly he had the nomination and a twenty-point polling lead on President Gerald Ford. The general election was close, but no matter: Jimmy Carter had just catapulted himself into the world's most powerful office on the strength of his personal virtue. At a time when the nation was sick of lying Washington lifers, Carter's pitch as an honest outsider perfectly painted the corner. Whatever was wrong with Washington could be fixed by what was right inside the heart and mind of Jimmy Carter. He established an approach that would later be copied by other outsiders seeking the nation's highest office: Ronald Reagan, Bill Clinton, Barack Obama, and Donald Trump all successfully used the Carter playbook. Reagan even used it to beat Carter.

Carter strode into the White House in January 1977, as the first occupant since Dwight Eisenhower to have never served in the U.S. Congress. Kennedy and Nixon both entered Congress together in the same 1947 "freshman class," Lyndon Johnson had been the master of the U.S. Senate, and Gerald Ford served as House Minority Leader for the Republican Party. All these men drew their understanding of power from Congress—its Byzantine rituals and unstated objectives, the power seniority and hierarchy wielded over every proceeding. Jimmy Carter did not care about any of that. He was a smart man, after all, and a moral man. Being untainted by Washington's corruption had just won him the presidency. Who were any of these legislators, neck-deep in the mire of America's national shame, to tell him anything? They should be looking up to him for leadership.

From the day of his inauguration Jimmy Carter provoked his own party's Congressional wing. They had just selected Thomas "Tip" O'Neill of Massachusetts as the Speaker of the House, only for the new president's staff to seat O'Neill's entourage in the topmost balcony while Carter took his oath. No oversight, this: When O'Neill's staff called to complain, the only response from the Carter camp was an offer to refund the tickets (members of Congress had to pay $25 per ticket, one of Carter's good-governance innovations). Dozens of staffers and vital allies never received a ticket. When selecting his team of confidants, Carter eschewed the Democratic establishment in favor of installing his personal friends and boosters from Georgia.[13]

To boost the flagging national economy, Carter proposed a "rebate" of $50 to every taxpayer nationwide. O'Neill and other Democratic leaders thought it a terrible policy, so broad that it would attract furious right-wing opposition while also failing to adequately boost consumer spending. Yet Carter refused to compromise, insisting that his research supported the rebate, and so those party leaders dutifully sold it to their members— only for Carter to abruptly change course and renounce the rebate with as little consultation as he had taken for the initial idea. Carter had been president for less than a year and already had his own party rending their garments. He wasn't trying to antagonize them; he genuinely felt his solutions were correct. That every step he took caused a metaphorical rake to spring from the metaphorical lawn grass and strike him in the forehead seemed not to impress him. Soon Carter would devise his greatest provocation, foreshadowed by his 1974 veto of the Flint River dam: a wholesale assault on what remained of the big-dam consensus.

In February, just a month after his inauguration, Carter worked himself into a froth reading a report his staff had prepared on the nation's water resources. Appalled at the inefficiency of recent dam projects and the questionable bookkeeping routinely practiced at both the Bureau of Reclamation and the Army Corps of Engineers, the new president decided to use the issue to discipline Congress. In his fashion Carter took a pile of reading materials into his office and made his decisions behind closed doors. He emerged with a list of roughly eighty dam projects around the country he wanted canceled, each singled out for its high costs and low benefits. Carter instructed his staff to surreptitiously pass the list around Capitol Hill so Congress would know he was serious. It worked; legislators from both parties reacted to the "hit list" with teeth-gnashing rage.

Winter snows had scorned the West that year, provoking an historic drought. Yet Jimmy Carter meant to cancel planned dams and water infrastructure for many of the states most impacted. California, Colorado, and Arizona all stood to lose projects. Politicians from those states were furious. As historian Rick Perlstein writes, "what looked like an inexplicable boondoggle often looked like a matter of life and death for the congressmen in whose districts those projects sat."[14]

Thus far Carter had relied on the news and television media to carry his messages directly to the American people. It played well enough when Carter was forcing fat-cat Congressmen to pay for their own inauguration tickets and selling the presidential yacht Richard Nixon had so enjoyed during his time in office. But dams and waterways were big business around the nation, were intimately connected with industries in every state, and protected countless millions against the loss of their lives and homes. The press declined to follow Carter's lead. David Broder, the establishmentarian *Washington Post* columnist, could scarcely believe "that Carter would let something like the Red River Project put him at odds with [legislators] whose cooperation is essential for passage of all the vital . . . legislation on the administration's agenda."[15] The Red River Project meant to spend $900 million constructing a barge-friendly waterway between Shreveport, Louisiana, and the Mississippi River 236 miles away. Senator Russell Long dubbed it "our birthright" and declared, "I think Red River will be funded, even if I'm not right about the economics."

On April 18 Carter made his final pronouncement: eighteen projects to be cut across a dozen states, including the staggeringly grandiose and expensive Central Arizona Project. Congress spit it right back in his face by passing an appropriations bill funding seventeen of the eighteen projects.

An angry White House intended to veto the bill but knew that dams and water projects were so popular that Congress would likely override his veto. They began lining up support behind the president—not enough to win a floor vote outright, but enough to hold the line in the House against the two-thirds majority required to override (they considered the

Senate vote already lost). Many of these Congressmen were Republicans otherwise opposed to the President's agenda. They demanded, and received, absolute assurance that if they stuck their necks out for Carter, he would veto the bill. Yet at the eleventh hour House Speaker Tip O'Neill managed to cut a deal with the White House. Half the projects would be approved and the other half cut, in addition to cutting a nuclear reactor project Carter opposed for environmental reasons. Carter caved and signed the bill after all, torching and embarrassing every single member of Congress who'd just taken on major risks to support him.

Congress humiliated the president once again the next fiscal year, when O'Neill attempted to fund all the projects he had agreed to cut the year before. He hadn't promised *never* to push these projects, after all, merely to omit them from the 1977 bill. Now Carter felt tricked, though, under the circumstances, he could scarcely complain. In retaliation he exercised his Constitutional prerogative to veto the entire 1978 appropriations bill. To shore up his position, Carter enlisted the help of Howard Jarvis, the right-wing businessman who had just helped bankroll an antitax revolt in California (Proposition 13, which capped property taxes at 1977 levels and crippled the state's finances thereafter). Jarvis attacked the appropriations bill as reckless government spending. Enough of Congress shied away from a fight with the swelling right wing to sustain Carter's veto. The notion of a Democratic president using the far right to attack public spending proposed by his own party would have been unthinkable just ten years earlier but, by 1978, the New Deal was truly dead, with Jimmy Carter gleefully shoveling earth into the grave. "Government cannot solve our problems, it can't set our goals, it cannot define our vision," he proclaimed during his State of the Union that year. Franklin Roosevelt had to be howling in the great beyond. If the government cannot do anything valuable, one might have asked Carter, why the hell did you run for president?

The Good Shit

For all Jimmy Carter's mishandling of Congress and his abandonment of four decades of Democratic Party orthodoxy, he was also a victim of circumstance. Inflation drove consumer prices higher by nearly 10 percent every year, even setting aside volatile commodities like food and fuel. Corporate profits were falling. As far back as the Nixon administration, policy elites had noted the increasingly zero-sum competition underway between companies and workers and suggested that one or the other was going to eventually suffer a harsh adjustment. America's executive class understood this better than the average citizen, and they were determined not to be left without a chair when the music stopped.

Painfully aware of the zero-sum game the U.S. economy had become, Carter did his best to escape the trap with policy. Established New Deal economists thought of inflation as an unfortunate byproduct of rapid growth; rapid inflation coexisting with a stagnant economy was supposed to be impossible. Franklin Roosevelt and Harry Truman had battled inflation with a combination of price controls, wage increases, and public investments targeted to specific sectors of the economy. But by the late 1970s, even Democrats rejected such hands-on policy. With left-wing economics discredited and flailing, policymakers turned to right-wing economists who offered tax cuts as a panacea for unlimited growth. They had an answer for inflation too, which they saw as a product of workers and governments spending on consumption, goods, and services rather than investment that might yield further profit. If wages were simply cut alongside the government spending that might replace those lost wages, then the remaining money would be more responsibly invested, and all would profit from the booming economy to follow. Desperate to boost America's declining prospects, Carter abandoned decades of New Deal economics in favor of tax cuts for wealthy people and large corporations.

Yet his interventions accomplished nothing, bulldozed by the crushing march of inflationary forces he felt increasingly powerless to control. His addresses to the American people, once a source of comfort that buoyed his poll numbers, took on an increasingly haggard and hectoring tone.

His policies, in other words, became watered-down versions of what the right wing of the Republican Party was already saying. The Republicans' path back to respectability after the searing disgrace of President Nixon lay in antitax animus. Both Jack Kemp and Ronald Reagan, the charismatic men competing to lead the party, built their appeal around warm smiles and bitter opposition to taxes of any kind. Reagan, the better salesman by virtue of being a truer believer, won out in the 1980 Republican presidential primary. In Jimmy Carter he would face an opponent who had deliberately hamstrung himself like no other sitting president in American history.

In August 1979, less than fifteen months from election day, Carter decided to shuffle his Cabinet. Having decided inflation was going to be his most dangerous foe going into the 1980 election, he meant to halt it at any cost. He replaced his Treasury secretary with the then-chairman of the Federal Reserve, G. William Miller. To fill Miller's position Carter appointed Paul Volcker, president of the fed's New York branch. Volcker made clear his plans for the office at their first meeting, telling Carter, "You have to understand if you appoint me, I favor a tighter policy than [Miller]."[16] Volcker was a deficit hawk and an inflation hawk, and, like Carter, he would pay any price to stop inflation and protect investors. Unlike Carter, he had an unlimited appetite for human suffering and no election to worry about.

Once given power, Volcker knew exactly what he would do with it: use the power of the Federal Reserve, which managed and regulated America's banking industry on a month-by-month basis through dozens of fine policy instruments, to deliberately crash the economy. Through suffering and unemployment, the system would be purged and made clean. He would do this by halting the use of all those instruments, paring the fed's tools back to the point where it would merely pace the rate of growth in the country's money supply, set interest rates, and sit back while the free market did its perfect work. Volcker set about instituting the most far-right-wing monetary policy of the twentieth century without any input whatsoever from the hundreds of millions of people who were about to suffer at his decree.

Volcker's fed colleagues pushed back, warning he might panic markets and cause millions of people to lose their jobs in the fallout. So be it, Volcker responded: the bank's mission now was fighting inflation, consequences be damned. "A pact with the devil," they called it, to which Volcker answered, "sometimes you have to deal with the devil."[17] When a reporter asked how much unemployment he was willing to tolerate, Volcker replied, "over time we have no choice but to deal with this inflationary situation." If the idols he worshipped demanded blood, blood they would have.

By early 1980 a recession had formally been declared, and while a summer of high job growth inspired a brief respite, at the first sign of inflation Volcker dropped the hammer once again. "The standard of living of the average American has to decline. I don't think you can escape that," he testified to Congress.[18] The year 1980 was a year of pain. "We will not try to wring inflation out of our economic system by pursuing policies designed to bring about a recession," Carter promised the American people as Volcker cranked the thumbscrews.[19] Yet Carter also refused his staff's exhortations to order Volcker to ease up, considering the intervention improper. Was it any wonder that a president who voluntarily kneecapped the country's economy and cast at least two million people out of work *in an election year* got demolished at the polls? Set aside Carter's other errors of judgment and temperament; his fate was sealed the moment he appointed Paul Volcker to run the Federal Reserve. Ronald Reagan, the former governor of California, grinned and fibbed his way through a campaign that ended with the incumbent president losing forty-four states.

Reagan was hardly a new breed of politician. At sixty-nine years of age, he would be the oldest president in history, and already behind the scenes he had begun to evidence the Alzheimer's Disease that would claim his life. The man was hardly a mental acrobat even in his youth—his first wife, Jane Wyman, had obtained a divorce in 1947, claiming unusual cruelty because Reagan made her attend his political gatherings and listen to his opinions. She referred to him as "diarrhea mouth."[20] Yet this middling actor turned union president turned governor turned far-right radio demagogue had an undeniable cunning and will. He had a stage actor's gift of reading a room, the finest shifts in its mood, and so little attachment to the truth that he could broadcast any message with the sincerity of a martyred saint. Every word that left his mouth he truly believed—he'd said it, after all.

Whereas Richard Nixon's guilty conscience sweated and labored through every lie and scheme, Ronald Reagan invited the nation to reside with him in an intoxicating fantasy. He pitched an inclusive vision of American greatness where any person could become wealthy, successful, and morally upright all at once so long as they worked hard and believed in themselves. None of his policies made any real inroads towards these goals. He cut taxes, wages, disempowered unions, demolished regulatory barriers, and reduced public investment to make way for private investment that would surely materialize if only the government got out of the way. It is no wonder that given two versions of the same right-wing program, the American people opted for Reagan's over Carter's. At least Reagan sounded optimistic about the future.

Needles and Dissecting Knives

President Reagan staffed his administration with a mix of loyalists from his tenure as the governor of California (from

1967 to 1975) and hardcore movement conservatives—the sort of far-right operators who once populated Richard Nixon's White House. In some cases they were the same people. George H.W. Bush, whom Nixon had elevated to national prominence, became Reagan's vice president. Patrick Buchanan, Nixon's press strategist, served as White House communications director from 1985 until 1987. Roger Stone, one of Nixon's henchmen, played a prominent role raising money and gathering support for Reagan's campaign and used that connection to drive his lobbying business once Reagan was elected. In 1987 Reagan would attempt to reward Nixon's Saturday Night Massacre hatchet man Robert Bork with a seat on the Supreme Court until the nomination failed. However chastened the Republican Party might have behaved whenever Nixon's name came up in public, behind closed doors the Nixon wing of the party experienced Reagan's triumph as their own reinstatement to power.

The new faces weren't much better. Reagan filled the government with people who were ideologically hostile to the idea of government, like Secretary of the Interior James Watt. Watt hailed from Wyoming, had worked as a Congressional aide on natural resources and, most recently, served as president of a right-wing law firm. At Interior he opened up huge tracts of public land to coal mining, oil drilling, and other environmentally destructive forms of extraction. It was no coincidence that oil companies had more than doubled their spending for the 1980 election cycle, kicking more than $25 million to largely Republican candidates.[21] Watt defunded or shuttered conservation programs at Interior, deregulated strip-mining companies, and called the resulting criticism "persecution" akin to what Jewish people suffered under the Nazis.[22] He tried to sell land donated for conservation purposes as mining tracts. He permitted coal and oil companies to drastically underbid for extraction rights, causing immense environmental damage while also shortchanging the American people whose interests he had sworn to serve.

Watt was operating from an established playbook right-wing provocateurs still use today: make offensive statements, denounce anyone who objects in similar terms, and paint oneself as a victim to court sympathy from the right's most rabid supporters. William Greider of *Rolling Stone* noted, "The interior secretary has fashioned a clever form of job security for himself: the more audacious he is, the more he appeals to right-wingers whom the administration can ill afford to offend. The man may be twisted, but he's not dumb."[23] These words could have been written about literally hundreds of Republican Party figures over the last fifty years. Watt was a brusque and bloodless figure, totally contemptuous of anyone less powerful than himself. "I have a Black, I have a woman, two Jews, and a cripple. And we have talent," he quipped of a commission he had appointed to facilitate coal mining.[24] The resulting outcry would ultimately prompt his resignation in 1983, demonstrating that Reagan's patience for bad press was not unlimited. He settled into a more comfortable role sniping at the left in the press, calling environmentalists "the greatest threat to the ecology of the West."[25]

To head the Environmental Protection Agency (EPA), Watt recommended to Reagan the agency's first female leader: Ann Gorsuch, a former corporate lawyer who had been swept into the Colorado state legislature by a tide of money from far-right beer magnate Peter Coors. Coors adored Reagan, generously funded his campaign, and had a number of his acolytes placed into the new administration by way of compensation. Gorsuch had no affection for the EPA or regard for its mission. Like Watt she set about undermining every function of the agency. She slashed her own budget, scuttled new regulations, and encouraged polluters to violate the existing ones. She purged the EPA of dissenting scientists and made life miserable for those who stayed.

Gorsuch ran into acute trouble with the Superfund program, established in 1980 to provide funding for the safe cleanup of locations tainted by hazardous substances. Evidence emerged that Gorsuch had directed the EPA to slow its payouts for work on three particular sites in California for political reasons. Congress investigated, demanded documentation from the EPA and ran headlong into the Reagan Administration's stubborn refusal to provide the documents. In December 1982 the U.S. House of Representatives voted to cite Gorsuch for contempt of Congress, the first time a Cabinet-level official ever faced such a charge.[26] The stalemate stood for months until EPA workers told journalists that Gorsuch had stated at a meeting on Reagan's yacht that she withheld $6 million slated for the Stringfellow Acid Pits waste site in order to avoid helping the U.S. Senate campaign of former California governor Jerry Brown.[27] This was a shot below the water line; the White House dropped their executive privilege claims and turned over the documents to Congress. Gorsuch resigned, though like most disgraced-and-resigned Republican officials she remained in the Party's good graces and was regarded as a sympathetic victim of Democratic perfidy. Her son, Neil, would go on to serve on the U.S. Supreme Court, elevated by President Donald Trump, who would gut the Superfund program much the way Neil's mother had.

Settling In for the End of History

Paul Volcker's shock therapy yielded the intended result: inflation gradually ebbed away as more than two million workers lost their jobs. Volcker's first recession, in 1980, had been sufficient to knock Jimmy Carter out of office. His second, from 1981 through the end of 1982, sent national unemployment to 10.8 percent. One could write an entire book about the human suffering caused by these policy decisions, but to Volcker it was all justified by one single number: the annual inflation rate, which fell from 14.76 percent in April 1980 and 3.85 percent in December 1982, and never rising above 5 percent for the rest of Reagan's presidency.[28] From the White House's perspective, the United States had emerged from a purging fire, withered but still vital and, most importantly, cleansed of inflationary sin. Reagan's boys had the opportunity of a lifetime, to remake the nation's political economy the way Franklin Roosevelt once had—meaning not just what the country produced, but how that production fed into the processes of power like campaign fundraising. Whereas Roosevelt based his New Deal on organized labor and the provision of public goods like social security checks and gigantic dams, Reagan trusted his new order to the finance and defense industries.

If you look at a chart of Wall Street's profits during the 1980s, it does not look like a boom era. The years leading up to Volcker's recession look much better, followed by a hard dip and then a gradual climb through Reagan's presidency. Relative to U.S. gross domestic product (GDP) it doesn't seem like much at all, and the Federal Reserve's tight-money policies limited the capital available to investors. Yet behind the scenes major changes were underway as Reagan's appointees aggressively deregulated the financial sector, essentially allowing banks to act more freely and dangerously as investors once the fed's spigot squeaked back open.

The defense industry drank deep from the public trough. President Reagan inherited a defense budget of $143 billion from President Carter, who had himself approved such large increases to that figure that it became a plank of Senator Ted Kennedy's 1980 presidential primary challenge against Carter. By 1987 the United States was spending $304 billion each year on a military that had not fought a genuine war in more than a decade.[29] The American war machine had never really wound down from the Vietnam War, never experienced a budget cut during the intervening years, and grew eager for opportunities to repair its damaged pride.

Reagan's administration pumped money not just into the official apparatus of the U.S. armed forces but also into its covert wings. The CIA and Department of Defense funded just about any overseas paramilitary group that declared itself anti-communist and did not worry about the consequences even as their proxies committed horrific human rights violations across Central and South America. They even funded the anti-Soviet adventures of a young Saudi expatriate named Osama bin Laden, whose later exploits will appear in the next chapter. Many of these right-wing killers even had CIA handlers assigned to them who oversaw their crimes and advised them, pocketing pallets of untraced American currency at every step of the way.

Much of that money found its way back to American soil in the accounts of private defense contractors and the companies who leased them out. The Iran-Contra scandal revealed the administration had been selling weapons illegally to the government of Iran in order to illegally procure different weapons for right-wing death squads in Central America, and eleven administration officials were convicted of crimes connected to the affair. Money poured into the defense industry even as Reagan slashed taxes on top earners, almost totally inverting the values of the New Deal coalition. The people who made up that coalition still existed, but Jimmy Carter's Democratic Party had already told them to look elsewhere for aid. No one with a platform would elevate their cause. Advocates for public investment no longer wielded any significant power in the United States.

On the opposite geopolitical pole, the Soviet Union limped toward collapse. Even a superpower needs allies; the Soviets needed partners aligned with the postcapitalist economy they wanted to build, yet the capitalist order that ruled the rest of the industrialized world had despised communists of every stripe since even before the Russian Revolution. World War II offered in its resolution a golden chance for the USSR (now led by Stalin), a golden opportunity to expand. Much of Eastern Europe, first overrun by the Wehrmacht and then reconquered by the Red Army, entered the Soviet orbit. Countries like Poland, Slovenia, and the former Czechoslovakia stood as military bulwarks against a potential invasion from Western Europe and also as the ideologically friendly trading partners the USSR needed to compete with the United States and its North Atlantic Treaty Organization (NATO) allies.

By the late 1980s Soviet leaders knew their system could not endure. The network they'd built was falling apart under the weight of its own contradictions and the ever-louder demands of people in the USSR's client states for full autonomy. Soviet President Mikhail Gorbachev tried to stabilize the system by announcing democratic and economic reforms, but it was all too little and too late. The Soviet Union formally dissolved over the latter half of 1991, and by December Gorbachev resigned. Billions of people who had accepted the Cold War as a back-drop of their lives breathed a sigh of relief. Somehow the two nuclear superpowers reached the end of their conflict with-out destroying the world.

Ronald Reagan retired to California after the 1988 election, replaced by his vice president, George H.W. Bush. President Bush continued the militaristic policies of his predecessor even as the media painted him an uptight bookkeeper. For decades the Pentagon had been checked by the need to avoid open con-flict with the Soviets, lest that conflict cause a nuclear war. Now they could send troops, spies, paramilitaries, or any combina-tion of these just about anywhere in the world without fear of reprisal. Bush experimented with this new freedom by launch-ing a six-week military operation against Iraq, in order to push its army out of neighboring Kuwait. The resulting "war" was a fabulous success for everyone involved on the American side, from military officers to journalists, and it might have pro-pelled Bush to a second term if not for an eight-month reces-sion in the United States from 1990 into 1991. The economy rebounded slowly, and the period was a sour memory in the minds of many voters in 1992.

Bill Clinton won that election, the first Democrat in more

than a decade to hold the White House. Clinton had no intention of going back to the New Deal era. "A government that offers more empowerment and less entitlement," he told the Democratic National Convention during the 1992 campaign, "more choices for young people in public schools and more choices for older people in long-term care. A government that is leaner, not meaner; that expands opportunity, not bureaucracy; that understands that jobs must come from growth in a vibrant and vital system of free enterprise." He offered Reagan's values in a kinder, more consumer-friendly package. Clinton officials proudly called this approach "triangulation."

Aiding their fortunes, the Federal Reserve chairman, Alan Greenspan, finally threw open the spigots the central bank had held almost completely shut since 1980. Easy money flooded the financial system. Investors attacked every opportunity in front of them, especially those in the newly opened Soviet Union. Crucial state-owned enterprises went up for auction across Eastern Europe; vultures eagerly descended on what remained of the USSR's empire. Wall Street rarely had it better than in the 1990s, and Americans have never seen that kind of decade-long boom since. Capitalism seemed to finally be elevating everybody, or close enough to everybody that the remainder could be written off as addicts and criminals. Incarceration rates, which had been skyrocketing since the early 1970s, climbed even faster under President Clinton.

American elites really did believe they had figured it all out: low taxes, low regulatory barriers, cheap money, and free-wheeling banks atop a massive military structure that represented the only remaining sector of the economy still receiving active public investment. If that military was unseemly due to its size and lack of restraint, it still served an extra purpose by keeping foreign governments pliant and open to American investment.

They started talking about "the end of history." The academic and writer Francis Fukuyama gets tarred as hubristic for coining the phrase, but he didn't mean that meaningful events would cease to occur. He meant that the question of how a society ought to be arranged and managed—to that point the central question of political philosophy—had finally been answered in an optimal way. We knew the answer now: a Western-style liberal democracy running a deregulated system of capitalism. President Clinton never explicitly endorsed this idea, but it is useful to understand the approach of American government under Clinton. The role of government was to create a nurturing environment for capitalism, at home and abroad, and otherwise get out of the way. Stability and growth, forever. Everything else was just details.

8

THE WAR ECONOMY

Just a few weeks into my senior year of high school, on a Tuesday, I was killing time during a free period. Walking through a cafeteria I spied my friend Alex jogging through in a state of agitation, as though someone was missing. He told me a plane had just crashed into the World Trade Center.

We retreated to the crew pit behind the auditorium stage, a hushed and cozy place with black-painted walls, and listened along with a gathering assembly as the radio attempted to describe the scene. Surely this was some awful accident, we murmured to each other: human or mechanical error ending in the worst imaginable outcome. Amid this the anchor's patter arrested suddenly, mid-sentence, then restarted with a clarity that in retrospect was impressive. He informed us, in a measured and professional way, that a second airplane had just struck the other tower. The obvious inferences he left unstated, but they fell on all of us with physical weight. It felt like watching a mountain walk, or the ocean recede before a tidal wave. Something too big to move simply *moved*.

We processed it together, there in the crew pit: a dozen teenagers clustered around a two-speaker stereo, afraid to talk lest we miss the next phase of the calamity unfolding. No adults present, but what could they possibly have told us? Eventually the school bell rang out the next period, and we filtered out of the pit into the brightly lit corridor of another world. They sent us home early, at least. It was already clear that there was going to be a war.

George W. Bush's presidency had, to that point, been characterized by verbal flubs most of the public found amusing. So prosperous had been the 1990s that the central issue of the 2000 presidential campaign became how to spend the federal budget surplus. The entire election season felt eerily superficial even by American standards, and the Supreme Court capped it off by peremptorily ending Florida's recount, handing the election to the Republican candidate by a 5–4 party line vote while declaring their decision could never be cited as precedent. The Democrats meekly conceded. After all, it did not really matter who led the country. Communism was vanquished. Free markets governed the Earth.

It was under these uneasy circumstances that George W. Bush, nine months into his first term, suddenly found himself a wartime president. After seven minutes frozen in panic before an audience of schoolchildren, clutching a picture book, he rushed to Air Force One while his staff prepared to present him to a shattered nation. Soon we had troops in Afghanistan, attempting to crush the terror network that planned the 9/11 attacks. In 2003 we invaded Iraq on transparently false pretenses, killing at least 500,000 and as many as a million Iraqi civilians.

For all the work done by Republicans and Democrats alike to undermine notions of collective enterprise, in the year following the 9/11 attacks people were desperate to contribute. Whatever it took to restore our sense of security we were willing to do. If the government had said, "this is the time to decarbonize our economy," we would have buckled up and made the necessary sacrifices. We needed only direction. Instead, President Bush told us the most important thing we could possibly do was to invest in the flagging economy. Go shopping, he said, while we mete out violence in your name. This once-in-a-generation impetus for collective action was squandered, boosting a few lousy quarters of retail.

Governing from a war footing kicked the listless Bush administration into gear. They cast for any element of the federal government's monumental portfolio that might possibly be threatened, and high on their list was the nation's water infrastructure. Bush officials imagined any number of gruesome scenarios that might result from a 9/11-scale attack on a major dam. Countless American cities lie in flood plains protected by dams, so federal authorities moved swiftly to lock down dams across the country. Unfortunately, many dams also serve as important road bridges (the scenic stretch of U.S. Route 93 that formerly ran over the Hoover Dam is a classic example), and the imagined threat of vehicular bombs caused the sudden closure of these roads. As no specific threat to any dam ever emerged, the nebulous condition of lockdown could never be relaxed. As of 2010 Route 93 runs across a separate bridge bypassing the dam. That bridge is named jointly for former Arizona governor Mike O'Callaghan and deceased Arizona football legend Pat Tillman, who died from friendly fire in Afghanistan while serving in the U.S. Army.

At the Folsom Dam in California, featured earlier in this work, the Bureau of Reclamation shut down the dam's roadway over which passed eighteen thousand vehicles every day. Anyone attempting to cross the American River had to do so over a narrow bridge downriver, built in 1918. Despite pleas from the locals to reopen the road and relieve the awful traffic caused by the closure, the bureau refused. Years later, in 2007, the Bush administration finally allocated nearly $50 million to build a new road bridge in the shadow of the dam. Similar scenes played out across the country, and often the easiest answer was to write a check. There was no limit to the nation's resources, provided you could claim a halfway plausible link with national security.

All Hazards: The History of FEMA

For nearly all of American history, the federal government has involved itself in disaster response. The member states are limited in their individual resources, especially during emergencies that constrict commerce and tax revenue. From the founding they made a regular habit of appealing to Congress after natural disasters, and Congress generally responded with a bill specific to each calamity. Between 1803 and 1950, they passed 128 such bills. The Federal Disaster Relief Act of 1950 streamlined the process, investing the president with authority to disburse federal emergency funds to the states without Congressional approval, provided a state's governor formally requested aid and the president declared a major disaster.

The year 1968 saw the National Flood Insurance Act (NFIA), mandating the adoption and enforcement of flood-resistant construction on top of liability insurance for properties in designated "Special Flood Hazard Areas." This came in response to a series of major hurricanes in the east and southeast, causing catastrophic damage and many deaths. The NFIA spread out the costs of reconstruction, and it encouraged architectural changes that limited future damage, but another punishing sequence of hurricanes and tornadoes in the 1970s spurred the Carter administration to consolidate every mechanism of federal disaster response into the new Federal Emergency Management Agency (FEMA) in 1979.

Carter appointed as FEMA's first chief a man named John Macy, who did by all accounts an excellent job convincing the many subagencies of his sprawling organization to pull in the same direction. Macy taught his subordinates that no matter the type or source of a disaster, the response should be much the same: rescue people in active danger and provide material comfort for everyone else. Whether facing hurricanes, earthquakes, nuclear radiation leaks, or volcanic eruptions, FEMA found the crucial tasks were providing people with shelter, food, fresh water, and ice. No matter the scenario, the most important thing was to have these critical supplies on-hand alongside an armada of trucks to deliver them.

President Reagan's administration didn't care about any

of that. They had the Cold War to win. Rather than appoint a career manager to run FEMA, in 1981 Reagan picked Louis Giuffrida, a former army colonel and an ally from Reagan's gubernatorial tenure in California. Giuffrida was the kind of right-wing figure who would have fit perfectly in the twenty-first century. Unflinchingly loyal to his boss, he was also irascible, condescending, and vain. He wore his hair lush and schoolboy-low like Reagan's, insisted on being called "General," and wore a military uniform complete with sidearm each day at his decidedly civilian office job. While serving in the U.S. Army just a decade prior, Giuffrida had written an academic paper titled, "National Survival—Racial Imperative," laying out a strategy to imprison nearly every Black American in concentration camps.[1] Giuffrida died in 2012, a form of cosmic justice sustaining him long enough to see President Obama serve a full term.

Giuffrida's tenure was every bit the ideologically driven circus one might expect given his background. He directed FEMA's funding and energy toward such projects as studying whether factory workers could survive nuclear explosions by jumping into large cisterns of water kept on-site for the purpose. He also directed the creation of a program to protect the leadership of the United States during a nuclear war by driving them around the country in unmarked vans loaded with communications gear. Naturally, decoy vans would be necessary to throw Boris Badenov off the scent and, before long, the United States was paying $160 million per year to operate a giant fleet of black vans whose only practical purpose was fueling conspiracy theories. Giuffrida continued to enjoy the president's favor until 1985, when it was revealed he had used $170,000 in public money to build himself a house in exurban Maryland, complete with a $29,000 luxury kitchen. He resigned in disgrace and, for the rest of Reagan and George H.W. Bush's time in office, the FEMA directorship was a patronage office tossed to minor cronies.

Bill Clinton's presidency saw the agency rebuilt under the

guidance of Clinton confidante James Lee Witt, who had managed floods and tornadoes at the county level in Arkansas and did, by all accounts, an excellent job with FEMA. Witt came to the position armed with fresh funding and a keen understanding of the needs of ground-level responders, but his most valuable asset was his close relationship with the president. At any time of day or night, Witt could call the White House and be put through to Clinton. He knew it, and everyone else in the government did, too. When Witt's office needed anything, be it pallets of bottled water or military airlifts, he got it. During Clinton's first year in office, the Mississippi repeatedly overran its banks and levees, causing heavy flooding in several states, and, in January 1994, Los Angeles suffered the Northridge earthquake. Clinton declared a record fifty-eight federal disasters during his first year, thrusting FEMA into view all over the country.

In April 1995, domestic right-wing terrorists bombed the Alfred P. Murrah Federal Building in Oklahoma City, killing 168 people. In the aftermath, a Congressional panel recommended FEMA incorporate counterterrorism operations, but Witt resisted. He knew FEMA's history and thought its disaster relief mission would suffer if it chased nebulous national security threats. Witt's lieutenants pushed back, arguing that if these contracts and responsibilities were being handed out then FEMA should take their share. In the end, counterterrorism remained in the Department of Justice's portfolio. President Clinton bequeathed to his successor a well-funded FEMA operating at the top of its game.

On the surface the incoming administration appeared to hold the right priorities: Bush had praised Witt and FEMA during his campaign as examples of government done right and nominated Joe Allbaugh—his close friend and campaign whipcracker—to run the agency, maintaining the close proximity between FEMA and the White House that had served the country so well. In truth the Republicans resented any federal agency that distinguished itself with competence. FEMA's success cut frustratingly against their ideological project of shrinking and dismantling government, so Allbaugh set about unmaking the practices that made his office effective.

Chief among these were mitigation grants bequeathed to the states in order to help them limit the damage of future disasters. In practice that money went toward all sorts of things: purchasing generators, widening roads, or sometimes projects the states wanted to fund off-budget. Clinton and Witt didn't much care about the bookkeeping. They reasoned that state budgets were always strained after disasters (unlike the federal government, they cannot print money) and any relief money would probably be put to good use. Allbaugh was not having it. He cut the federal subsidy for postdisaster mitigation from 75 percent to 25 percent, claiming he would make up for it with predisaster mitigation grants.

This was a neophyte's view, the opinion of a man who hadn't studied the field and did not have the sense to heed what every expert in FEMA told him: valuable as premitigation might be, postmitigation could not be cut without exposing cities and states to vastly more risk over the long term. Disasters reveal vulnerabilities like nothing else, and if the feds did not fund that work, it simply would not get done. Allbaugh didn't care. He implemented his agenda while ousting officials he deemed too close to Clinton or Witt. He went on television to blame the town of Davenport, Iowa (freshly inundated by Mississippi River water) for poor planning. The people of Davenport were the authors of their own misery, Allbaugh declared. "I don't know whether it's two strikes . . . three strikes, you're out. But obviously these homes and properties that are continually flooding, it's not fair to the American taxpayer to ask them time in and time out to pay for rebuilding."[2] At the same time, Allbaugh was agitating for FEMA to have a more active role in national security policy.

He got his wish, in a way. The September 11 attacks created a major push to consolidate all federal domestic security agencies, and the result was the new Department of Homeland Security (DHS). Senior policymakers mandated that the DHS have disaster-response capability. As its first secretary, Tom Ridge, said, "If you didn't have a FEMA at Homeland Security, you'd have to create one. Its mission of response and recovery is at the very core of what we do."[3] So into the big pot went FEMA. Allbaugh protested bitterly that severing the agency's direct line to the president would cripple it but found that his voice counted for little. He resigned, telling his bosses they would regret their decision "sooner rather than later."[4] The man who would replace him at FEMA was a lawyer and former head of the International Arabian Horse Association with no experience in disaster response, the now-infamous Michael "Brownie" Brown.

Brown was a friend of Allbaugh's, first hired as FEMA's general counsel before Bush elevated him to the directorate in early 2003. Though disliked by many colleagues, Brown understood the core function of his agency was delivering material goods to disaster victims as quickly as possible. Yet from the DHS's perspective, FEMA was just one more tool in the national security state's rapidly expanding portfolio. No one thought much of it until catastrophe struck in the summer of 2005.

A Storm for a New Century

Hurricane Katrina first attracted the attention of North American meteorologists in the second week of August, over the Atlantic Ocean near the Bahamas. It earned a name on August 24, the eleventh storm of the year to earn that grim distinction, and, within hours, the National Hurricane Center issued warnings for the east coast of Florida. Katrina was not a big storm at that point, but she drove wind and water with surprising force. A quick eight-hour slash across the Sunshine State's midsection loosed thirteen inches of rain and dropped enough trees to kill six people. The storm drew fresh strength from the warm Gulf of Mexico, swelling, its winds now exceeding one hundred miles per hour, and officials predicted it might land anywhere from Louisiana to the Florida panhandle. Officials from five states rolled into action, all hoping the work they queasily undertook would turn out to be precautionary.

In Washington DC, career FEMA planners gnawed their cuticles. Katrina would be dangerous over any populated region, but New Orleans was especially vulnerable. Disaster experts have long regarded a major hurricane event in New Orleans as a worst-case scenario: built on a muddy spit of peninsula jutting into the Gulf and bounded by the brackish inland bay of Lake Pontchartrain, it is bounded by water on three sides. Much of the city lies below sea level, and shipping canals penetrate deep into its downtown region. New Orleans relies on a vast system of levees, seawalls, and pumping stations to keep it dry even in the best of times. It is an urban center created by infrastructure, its existence made possible only by feats of engineering that would have seemed supernatural to the city's Acadian founders. Storms routinely flood streets and neighborhoods; any longtime resident of New Orleans can tell stories of water rushing up their front steps. Given a sufficient storm surge, Lake Pontchartrain could potentially tip its contents like a teacup into the city.

The morning of August 27, FEMA officials laid out a potential catastrophe to the DHS. It was conceivable, they wrote, that Hurricane Katrina could drive a storm surge high enough to totally collapse New Orleans's flood protection and could kill sixty thousand people. This message never reached DHS secretary Michael Chertoff, who spent that day working from home on immigration issues. Michael Brown, for his part, said the right things and notified the White House (then operating remotely out of Bush's vacation property in Texas) that a potential disaster was imminent. Yet FEMA workers on the ground in Louisiana were already incensed over logistics. Journalists Christopher Cooper and Robert Block, in their excellent book (called *Disaster*) on FEMA's response to the Katrina tragedy, describe a career responder upbraiding Brown: "Where's my goddamn ice? Why can't you ever get me ice?" he snapped, before saying he'd acquire it himself and hung up.[5]

Several years before, FEMA commissioned a highly granular simulation of a catastrophic hurricane impact on New Orleans. Hurricane Pam was the fictional storm's moniker. The resulting study contained state-of-the-art contingency planning for such an event, but the agency had by all accounts ignored the Hurricane Pam report and now struggled to locate enough copies to brief its own leadership. FEMA positioned its standard hurricane supply allotment outside New Orleans but did not direct additional resources or personnel to the region.

On Sunday, August 28, FEMA staff assembled another alarming report for DHS warning them that Katrina was now category 4 and noting at least one hundred thousand New Orleans residents lacked the transportation to evacuate. If this made an impression on anyone it wasn't Michael Brown, who went on television later that morning to chirp optimistically about the response. A video conference including Brown, Bush, and Chertoff emphasized the seriousness of the storm but led to no adjustments or additional measures. Everyone seemed to agree that Katrina might cause significant property damage, especially along Lake Pontchartrain and the mouth of the Mississippi River, but they hoped New Orleans would be more or less okay. Never mentioned at that conference was a threat over which Mayor Ray Nagin and locals were already wringing their hands: that the storm surge would be sufficient not just to spill over the top of the city's flood levees but to breach or topple them completely.

Forty Years and a House of Cards

As the most flood-vulnerable major city in the United States, New Orleans has always gambled ambitiously with the elements. Natural levees adjacent to the Mississippi River sheltered settlements on that spit of land, much-trafficked even before colonization by Indigenous people moving goods between the river and the lake, which they named Ok-wata. The French built New Orleans in the early 1700s on a crescent of high ground overlooking the river's mouth and almost immediately started digging canals to shunt the Mississippi's legendary surges toward the lake. New residents built their homes on ever-lower land, draining the swamps earlier generations had used as a sewer. The French ceded their city to Spain at the end of the Seven Years' War in 1763, before Napoleon won it back by treaty in 1800 and sold it to the United States in 1803, but every regime faced the same relentless encroachment of water. Flooding was a seasonal certainty, and anyone who wanted a building to last more than a few years had to raise it on several feet of struts. Many families ringed their property with hand-shoveled dirt levees and hoped for the best.

The Industrial Revolution expanded Americans' engineering ambitions by an order of magnitude, and for the first time Americans imagined human ingenuity might put a stop to these crises. Bonds and property taxes funded the creation of the New Orleans Sewerage & Water Board, which drew up detailed topographical maps, improved the existing network of drainage canals, and installed pumping stations to hasten the water on its way.

The city raised levees to protect what used to be its own flood plain, in which many of its poorest citizens now resided. Yet the design of these works was fundamentally flawed, as a city surveyor named W. H. Bell insisted in 1871 (and was ignored): under certain conditions they could actually trap and magnify a storm surge. This happened in 1915 and again in 1927, at the cost of hundreds of lives. World War II made the port city more important than ever. By 1945 all the land between the Mississippi and Lake Pontchartrain was drained and developed. Each time a storm wrecked the flawed flood walls, the city built them higher.

Hurricane Betsy proved in September 1965 that even those

systems were not enough. Though most of the city avoided serious flooding, the storm surge breached levees protecting New Orleans's Lower Ninth Ward. Eighty-one people died as a direct result of the hurricane, and many others likely perished from exposure and deprivation in its aftermath. The federal government stepped in with the city's assent, assigning the Army Corps of Engineers ultimate responsibility for financing and maintaining New Orleans's levee system.

The corps embarked on a massive project to upgrade the city's flood protection with the goal of withstanding a "hundred year" storm—that is, the worst storm one might statistically expect in the span of a century. They spent hundreds of millions of dollars pursuing that worthy mark. When Katrina landed forty years later, the work still was not finished, but they still had a great deal to show for it: more than one hundred miles of hurricane protection, including thirty miles of walled river levees and two hundred flood gates. The corps operates hundreds of miles of flood protections across coastal Louisiana, as well as several inland shipping canals. One in particular, the Mississippi River Gulf Outlet (MRGO), shaved forty miles off the route from downtown New Orleans to the Gulf of Mexico but destroyed a vast swath of marshland in the process. Coastal marshes are smelly and often unsightly parcels of land, but they are the best flood buffers on Earth. When surges hit, marshes act like sponges, absorbing millions of gallons of storm-tossed water. In this instance and others, the corps's shipping mission stands in direct conflict with its flood protection mission. That's especially true once one considers the considerable funding drawn by projects like MRGO, which serves very few ships and must constantly be dredged so it does not silt up, against the corps's protests of penury when it comes to flood control in New Orleans.

Due to its watery nature, Louisiana has long been a net beneficiary of the federal government. The state's Congressional delegation, especially its U.S. senators, found their ways onto influential committees and directed funding back home. By the mid-2000s, with the War on Terror in full swing in multiple theaters, the Bush administration was gleefully slashing domestic spending wherever they could. In 2003 New Orleans requested $11 million for flood control; Bush offered $3 million, which Congress increased to $5.5 million. In 2004, they asked for $22.5 million; Bush offered $3.9 million, and Congress once again bumped it to $5.5 million. In 2005, the year of Katrina, Bush offered $3 million to protect New Orleans while allocating $14 million merely to dredge the MRGO.[6]

Protecting the residents of New Orleans just was not a major priority for the federal government. Perhaps this reflected the country's political drift rightward, emphasizing the needs of big businesses like those who sent barges down the Mississippi. Perhaps it reflected a sense of security bred by good luck: in those forty years between Betsy and Katrina, New Orleans suffered no major hurricane strikes. In either case, the "least cost" principle on which the corps operated turned out to yield more costs down the line.

Flood walls built by the Corps of Engineers adhere to three basic designs. The first and most basic is the earthen levee: just a big heap of clay and dirt that holds back water very securely but occupies a huge amount of acreage and is thus impractical for protecting established neighborhoods like those in New Orleans. There are "T-walls," vertical metal plates anchored by wide bases that yield an inverted T shape with diagonal struts to buttress the wall. They are very secure in all conditions and occupy less space than a levee but are expensive to place. Last there are "I-walls," the same metal plates but without the base and struts. I-wall is the cheapest wall with the smallest footprint to boot, but it is only as secure as the soil in which it's sunk. If the soil moves, the wall moves with it.

New Orleans offers examples of all three designs among its copious flood protection works, but about thirty-five miles of it is I-wall. Most notably, the section of the 17th Street Canal that collapsed during Katrina and flooded much of the northwestern part of the city was I-wall. The project had been built shoddily and on the cheap, though Cooper and Block suggest blame should reach further than the corps: "the Orleans Levee District originally selected T-wall, but it balked when it saw the $44 million estimated cost. The same job constructed with I-wall would cost about $21 million. . . . 'They cut every corner,' said Fred Young, a Corps supervising engineer, of the local agencies' work."[7]

The London Avenue Canal was built with the same I-wall design as 17th Street. The Orleans Canal, elevated some twenty feet above the neighborhood it protects, had two conspicuous gaps in its walls where the work went unfinished for years on end. Bill Clinton's late-term drive for austerity and George W. Bush's obsession with terrorism left the city unable to defend itself. As Cooper and Block write, "In 2005, exactly forty years after it had been authorized to proceed, the Corps was only about 75 percent done with its work to protect New Orleans from the storm surge of a killer hurricane."[8]

Worsening matters, what work had been done fell scandalously short of basic engineering standards. Not only had the Corps of Engineers relied on I-wall for much of the region's protection, trusting in untrustworthy muddy soil, but those walls hadn't even been sunk to the depth prescribed. The corps insisted they had placed the walls at seventeen feet of depth, but subsequent investigations would reveal that they were sunk to only ten feet *in the corps's own diagrams.* The armor New Orleans relied upon for its survival was riddled with holes, and water is exceptionally good at exploiting weakness.

The Stinging Rain

On Monday, August 29, now at full category 5 force, Katrina made landfall. Most Bush administration officials had taken the president's Texas vacation as a cue to abscond from DC themselves. Leading the White House's response was Kenneth Rapuano, a nuclear proliferation "expert" whose normal job involved combing the ruins of Saddam Hussein's Iraq in search of weapons that turned out not to exist.

When Katrina finally hit the city, one of the first facilities submerged was the southern armory for the Louisiana National Guard. An entire fleet of high-water personnel transports were inundated and destroyed. Guardsmen fell back to their rallying point at the New Orleans Superdome, a monolith whose solid concrete construction and curved roof made it the city's sturdiest redoubt against winds exceeding 175 miles per hour. Some five hundred National Guardsmen took shelter there along with about nine thousand city residents, though that second number would swell to twenty-five thousand in the following days as rescued civilians were deposited at the dome. The Morial Convention Center, just uphill from the Mississippi River, also became a major evacuation site, but it lacked the Superdome's stockpiles of food, water, and medicine, and conditions quickly began to deteriorate.

In the morning hours of August 29, as light came up through a heavy pall of clouds and stinging rain, residents of the Ninth Ward heard a loud *boom*, or a series of booms in quick succession. Citizens investigated only to discover several feet of flood water advancing down the streets. The worst-case scenario had suddenly come to pass: the levees were breached, and the contents of the Industrial Canal were emptying into their neighborhood.

The 17th Street Canal went next, inundating the Seventeenth Ward. In contemporaneous accounts residents of the Ninth and Seventeenth described the sound of the fracturing levee walls as resembling that of a bomb. A few locals believe to this day that authorities deliberately demolished the floodwall with explosives, the objective being to flood a working-class neighborhood to save wealthier districts.[9] That is not how New Orleans's flood control system operates, and no hard evidence has ever emerged to suggest such a crime took place, but if you have read this far in this book you can understand why working-class Southerners might not trust their own government.

Levee breaches continued throughout the day while anyone who might have plugged them sat grounded by the hurricane. Water flooded the porous peat soil underlying the canals, gaining enough purchase to grip and move entire blocks of earth along with the I-wall barriers dug into them. By nightfall at least three-quarters of the city were underwater. In many places the water was so high that people had to climb away, first upstairs and then to the attic and finally to the roof. Spike Lee's moving documentary film *When the Levees Broke* plays 911 call audio from frantic people trapped in their attics by waters that were already at their neck; knowing they haven't the tools or strength to break through the roof, they beg operators for aid. Elderly and disabled people drowned terrified and alone in their homes. Others died in the suffocating heatwave that followed, their fans and air conditioners rendered useless for lack of electricity.

With their homes flooded and their power knocked out, people camped on their roofs seeking shade and hoping for aid. The U.S. Coast Guard worked around the clock pulling stranded people to safety, alongside U.S. Fish and Wildlife agents, New Orleans police, and bands of altruistically minded citizens in private boats. The "Cajun Navy" became folk heroes that awful week, where helicopters from CNN were more common sights in the sky than those from the National Guard or the Army. *Where are the feds?* people asked each other. The fact of the disaster they could accept, but everyone expected the government to show up forthwith, and the help did not come. Fittingly, just that summer the Bush administration had refused to fund a fleet of aluminum flatboats for search and rescue in New Orleans because the request was unrelated to terrorist threats. FEMA's supplies were inadequate even by the standards of their own plan, and their medical teams were not yet in place when the hurricane struck.

The federal government could have brought the full weight of their resources to bear on the unfolding disaster—they could have brought in military helicopters to rescue people, they could have ordered the Department of Transportation to rush evacuation buses to New Orleans, they could have done so many things. The rubric and authority for all of it was written in the National Response Plan approved by DHS just four months earlier, which gave the feds tremendous flexibility to respond to disasters. The problem was that Bush's appointees wrote the plan to distinguish between ordinary disasters (where it assumed local/state governments would lead with FEMA as a backstop) and "catastrophes" (where federal agencies would actively push every type of aid into affected regions and sort out the accounting later). Catastrophes were implicitly assumed to be connected with terrorism or national security, and thus the leadership at DHS fought to keep Katrina from being declared a catastrophe. They thought of military threats as being more important than natural disasters, demanding special designation, and wanted to avoid setting a precedent by urgently expending resources on something as trivial as a hurricane.

Compounding matters were the shortcomings of the man commanding the Homeland Security Operations Center (HSOC), through which all information in the massive department flowed. Brigadier General Matthew Broderick (not that one) made his bones in the Marine Corps and took a military approach to fact-gathering. He spoke often of the "fog of war," a military concept meant to account for both the practical difficulty of information-gathering in combat and the active misdirection expected from one's enemies.[10] Military reports focus on that which is conclusively known and omit anything else. Whatever merits Broderick's approach might have had on a battlefield, it was poorly suited for natural disasters. Misinformation might abound due to ignorance or confusion, but nobody in a hurricane is lying to first responders, and certainly no one is shooting at them. As reports of breached levees and rising waters filtered into FEMA, Broderick's HSOC simply discarded them as unconfirmed rumor—even if multiple sources corroborated them, *even if the reports came from officials who reported directly to HSOC.*

With their own command center junking reports, DHS and the White House had no idea how bad things had gotten in New Orleans or how quickly. Especially maddening in retrospect is the debate between public officials as to whether the flood walls had been breached or merely overtopped. Civilians and responders alike submitted scores of reports describing the breaches, including a twenty-foot chasm in the wall of the 17th Street Canal early Monday morning, and yet as the sun set that terrible afternoon, DHS was still unconvinced. In fact, much of the leadership would later tell Congress that they were under the opposite impression: that the walls had specifically NOT been breached. HSOC staffers on the ground came from military backgrounds and were unfamiliar with the city's complex topography, so they struggled to comprehend what civilians told them. Meanwhile, flood water soaked through the urban substrate to drown underground cables and shut down cellular and satellite phone towers. What few roads remained dry were blocked with debris. By the time Mayor Ray Nagin learned the levees had in fact been breached, that twenty-foot gap in the 17th Street Canal was a hundred feet wide. If there had ever been a window to "save" New Orleans, it was now closed. Only mitigation remained.

Broderick received nine reports of breaches by 5:00 p.m., but the report HSOC distributed over an hour later stated, "Preliminary reports indicate the levees have not been breached."[11] Broderick received about seven hundred emails on the Katrina situation that day but never opened any of them as he did not use email. He also did not read the New Orleans *Times-Picayune*, which reported the breaches as fact on the day they occurred. That evening, Michael Brown went on CNN to say, "We have some, I'm not going to call them breaches, but we

The War Economy

have some areas where the lake and the rivers are continuing to spill over."[12] Information was not moving up the chain of command, and the consequence was a total lack of urgency in the federal response. As far as they were concerned, Katrina was a run-of-the-mill hurricane event. President Bush spent the day on a jaunt to Arizona to celebrate the birthday of Senator John McCain, a man Bush's campaign once suggested fathered his Bangladesh-born adopted daughter out of wedlock. The two loathed each other but shared some cake for the cameras.

All of which is not to absolve state and local officials for their various missteps, well chronicled in the extensive reporting on Katrina. Mayor Ray Nagin failed to communicate important information up the chain of authority, and, as of this writing, is serving a prison sentence for wire fraud, bribery, and money laundering (before and after Katrina). Governor Kathleen Blanco husbanded her authority tightly at moments when only explicit requests for federal aid could have opened the proverbial spigots. The chief of police went on TV to describe humanitarian horrors, like murders and rapes at the Superdome that simply never occurred. Miscommunications occurred between local and federal officials in the field, delaying the Army Corps's response to the levee breaches. A catastrophe of this scale yields more than enough blame to spread around. I choose to focus on the errors of the federal government for two reasons: first, the Army Corps of Engineers bore ultimate responsibility for the construction and maintenance of the city's flood protection, and, second, the feds have increasingly backstopped state and local governments during disasters over the last century. We have built a massive centralized state that is liable both for what it does and does not do.

The Superdome

By the morning of Tuesday, August 30, the White House started to appreciate the situation's gravity. Michael Brown told Bush, Chertoff, and Vice President Dick Cheney that 90 percent of New Orleans residents had been displaced, and FEMA's resources were insufficient to the disaster. The upper levels of federal management shuddered into motion as Bush cut short his planned five-week vacation and prepared to return to Washington. From his inland headquarters in Baton Rouge, Brown tried to direct buses to handle the swelling tide of storm refugees. The sun came out, and the wind surrendered to an oppressive heat that drove those camping on rooftops to seek shelter upon higher ground. Through waist- or chest-deep water they waded and swam, sometimes with children piled into hollowed-out refrigerators. Some made for the Superdome and Convention Center while others congregated at the crests of highway onramps. Rescue teams deposited their charges on these concrete islands and departed to continue their work, assuming FEMA and state officials would arrange their evacuation. In many cases that help never came. Hundreds of people spent days marooned on sunblasted strips of highway without adequate food, water, or medical attention. When sick and elderly people died of exposure, their relatives did their best to keep the corpses shaded.

It was the evening of the thirtieth when Michael Chertoff finally declared Katrina an "incident of national significance," yet another nebulous designation in DHS's National Response Plan.[13] The feds still did not have a clear ground-level picture of the disaster, relying on occasional reports radioed in by armed DHS teams reconnoitering New Orleans like it was enemy territory. FEMA communication trucks sat idle, convinced the city was impassible even as supply trucks from the state government rolled in along a few dry roads. The teams underestimated the population sheltering in the Superdome by half. Since Broderick's HSOC was the only repository of information, the former marine general took it upon himself to direct ad hoc rescue efforts. He would see someone in need on CNN and attempt to order a rescue, only to be informed the footage dated from the day before. He badgered his subordinates about trivial discrepancies in figures. The federal effort was already maddeningly slow and inept, and Broderick's flailing only snarled it further.

Governor Blanco received assurances in the predawn hours of August 30 that FEMA was sending five hundred buses to evacuate the twenty-five thousand people in the Superdome, to arrive the following day. The vast arena's roof had leaked during the storm and its plumbing was flooded, so this host of humanity suffered under worsening conditions for several days. Food and water ran out. Scuffles occurred between frustrated citizens, but the refugees seem to have handled a terrifying and miserable ordeal with relative aplomb. With the Superdome's facilities on the verge of collapse, they were eager to leave.

The Department of Transportation assured the White House at 11:00 a.m. on Wednesday, August 31, that the evacuation would soon commence, yet five hours later the buses had not appeared. Blanco's people scrambled to find them but came up with nothing. In fact, FEMA's request had gotten gummed up in DHS bureaucracy. The buses were at least a full day away, but nobody knew this for certain or could communicate it to responders on the ground. The Superdome would not evacuate until Friday, September 2, by which time DHS still had not figured out a coherent plan for sorting and routing them. With every in-state resource exhausted, refugees were routed to Houston, to San Antonio, to Atlanta, to Kansas City—anywhere they might have an uncle, cousin, or schoolmate willing to lend shelter. DHS scrambled commercial airliners to get refugees out of town, only to have that process turn into an agonizing slog when the Transportation Safety Administration insisted on full security screenings for every single person through the gates. The War on Terror never slept, not even at that dark hour.

The U.S. Army arrived on the scene Wednesday night: the First Army, a readiness, training, and disaster-response command then based at Fort Gillem, Georgia. Leading the First was General Russell Honoré, a gruff and authoritative Louisianan who walked into an absolute mess with minimal advance intelligence. What reports they had, from their own people as well as the scenes on CNN, painted a cinematic portrait of lawlessness. Looting, murder, and rape were supposedly rife in streets surrendered by police to bands of heavily armed brigands. The Convention Center, the Pentagon had heard, was a veritable fortress that might require a thousand specially trained troops to crack.

None of these things were true, but Honoré did not know that. His first mission was to restore order and take back the city. Army and National Guard troops fanned out through the downtown neighborhoods expecting swarms of looters, to find nothing but exhausted and beleaguered Americans in need of food and fresh water. An officer riding a Humvee through town smelled chicken barbecuing nearby: "I realized right away that we weren't going to have any problems here."[14] They emptied the Convention Center without a fight, recovering nothing but kitchen knives and a single rusty handgun.

The Army's weapons were better maintained but similarly useless under the circumstances. Their olive-camouflaged arrival empowered a single man—Honoré—to take control of the situation. Bickering paused between local, state, and federal officials as they could all look to Honoré for the last word. He commanded the helicopters, the high-water trucks, the thousands of troops belatedly entering New Orleans, filling the vacuum FEMA had left. With the Army's backing, resources flowed into the area that should have been in place all along. Where DHS had dithered and bargained itself into a submissive posture, the Pentagon attacked the disaster with blunt instruments. Which, it turns out, is exactly what disasters demand: dump resources into the affected region and bill the feds. Unfortunately, rightwing fanaticism for dismantling public goods and a bipartisan obsession with terrorism led Bush to hamstring the civilian government's capacity to respond to disasters. When those agencies failed, as they were meant to, the only federal department with the no-strings-attached funding and latitude to step in was Defense. The U.S. Army and the Louisiana National Guard saved what was left of New Orleans because the rest of the government fundamentally lacked the capacity to do so.

Maneuvering in Floodwaters

On the surface, things in New Orleans were looking better. Behind the scenes jostling was already underway to dictate the aftermath. Bush leaned hard on Governor Blanco to sign over all relief efforts to the White House, promising more aid through the military, but she balked at surrendering control of the national guard. It seemed to Blanco (a Democrat) that Bush, having thus far shirked responsibility for the disaster, was now trying to take credit just as light twinkled at the end of the proverbial tunnel. Despite tremendous pressure from federal officials, Blanco's people held firm. "The specific question was, 'How does this give us one additional Federal asset?' And they never convinced me it would give us anything," recalled Blanco's chief of staff, Andy Kopplin.[15] The federalization order was never signed. It later came out that Bush had made the same proposition to Mississippi governor Haley Barbour (a Republican), who rejected it even faster than Blanco.

The White House was slow, fatally slow, to recognize the disaster unfolding in New Orleans. Once they did react, their consistent impulse was to manage public relations rather than making provisions for the victims. Only national security and terrorism really interested them; everything else became a question of perception, to be manipulated by media and state messaging. Especially important was the medium of television. President Bush, jolted out of complacency by the calamity unfolding on television at his Texas vacation property, sought not to contain the disaster but rather to change what the television was saying.

CNN cameras captured him frowning out the window of *Air Force One* as it cruised over the ruined New Orleans. Having cut his five-week vacation four days short, he stood in the White House Rose Garden on Wednesday, August 31, and rattled off a sequence of statistics purporting to show the government had adequate resources in place. On Thursday, September 1, Michael Chertoff insisted the Superdome and Convention Center were adequately provisioned for the disaster when this was obviously untrue. Michael Brown claimed only five thousand people were sheltering in the Superdome (the actual number was twenty to thirty thousand) and said they were being fed, only to rescind that claim on television when Ted Koppel told him it was not true.

On September 2 Bush landed in the region for an in-person visit, the better to try and control the disaster narrative. Despite the hundreds dead and tens of thousands who had lost their homes, Bush focused his sympathy on one man: Republican senator Trent Lott, a lifelong ally of racists and segregationists whose mansion had been destroyed. "Out of the rubble of Trent Lott's house—he's lost his entire house—there's going to be a fantastic house," the president warbled. He extemporized about how well everything was going, before turning to an uneasy-looking Michael Brown and uttering the words that would forever attach the FEMA chief's name to infamy: "Brownie, you're doing a heck of a job."[16]

Once the White House realized just how badly DHS, FEMA, and the rest of the government had failed, they started distancing themselves from the whole enterprise—working much harder at this task than they had at mitigating hurricane damage. Bush started insisting he had taken Katrina seriously starting on the August 30. Chertoff cut Brown further and further out of the loop, taking on more authority he did not seem to want: "I'll tell you my philosophy, since I guess it's my responsibility now," he sighed at a reporter who asked whether DHS had overprioritized counterterrorism. "I think we have to plan on both, because I think they're both mutually reinforcing."[17] The White House was content for the entirely expendable Brown to take the fall. He would shuffle out the door on September 12, 2005, two weeks after Katrina landed. Meanwhile, the morgues in New Orleans were choking on corpses, and FEMA had nothing to offer when it came to retrieving or storing human bodies. Dead Americans would lie rotting like garbage for weeks in full view of their neighbors before anyone showed up to collect them.

With Brown gone and New Orleans still a disaster area for the foreseeable future, the Bush administration claimed the city had in fact suffered *two* disasters: the hurricane itself and the failure of the levees. They had adequately prepared for the hurricane, they argued. The real damage had come from the failed levees, which nobody (they argued) could have foreseen. "We didn't merely have the overflow," Chertoff explained. "We actually had a break in the wall. And I will tell you that really that perfect storm of combination of catastrophes exceeded the foresight of the planners and maybe anybody's foresight." "It was unforeseeable that a Category 4 hurricane would knock down a levee designed to withstand Category 3 winds?" one reporter asked Chertoff. He did not get an answer.[18]

This notion of Katrina as a double disaster, a sui generis event beyond the ken of any disaster planner became the Bush administration's basic defense of their performance. This argument fails for several reasons, including the Corps of Engineers' engineering errors. To my mind the strongest rebuttal is the underappreciated fact that Hurricane Katrina missed New Orleans.

That's not to minimize the horrors that occurred; it is to note that the storm could have been so much worse. High winds and rain lashed New Orleans for two days, but Katrina actually veered east shortly before making landfall, and, as a result, she visited her worst destruction on the coastal communities to the east of the city. Entire towns on the state of Mississippi's Gulf Shore were simply blown away by a storm surge that may have reached twenty-five feet in height and swept twelve miles inland. The devastation was unholy and total.

Good fortune spared New Orleans the worst. Twelve inches of rain fell on some parts of the city, but others got considerably less, and the twelve-foot storm surge off Lake Pontchartrain should not have overwhelmed the city's flood protection systems. The federal government spent hundreds of millions of dollars over forty years to build category 4 flood protection, only to see it fail under a category 3 pummeling. Some levee breaches, like that on the 17th Street Canal, occurred as many as four feet below design specifications. Bush's underlings tried to work themselves off the hook by claiming a unique calamity, but it seems clear any substantial hurricane would have caused similar breaches. The federal government simply lapsed in its duties: first when building the flood walls, then when maintaining them, and finally when the bill for that negligence came due.

All this was the price of systemic public divestment in the United States. The right-wing ideologues who saw disaster response, infrastructure, and public health as onerous distractions from the Cold War now used the War on Terror to keep their scams rolling. When given power, they immediately set about dismantling crucial elements of the state in order to shower public money on defense contractors with horrendous records of corruption. This is war profiteering. It has been the defining trait of right-wing governance for the last forty years. Even the Army Corps of Engineers' $14 billion investment in armoring New Orleans against future hurricanes has already gone sour: in 2019, just eleven months after completing the project, the corps announced that rising sea levels will render their work obsolete in as little as four years.[19] As the economy of the United States has shifted ever more toward the finance and defense industries that make obscene profits off each others' malfeasance, every other element of our society has withered on the vine.

9

INUNDATION

The most important thing to know about the 2008 financial crash is that it was made up. Everything that followed—the human suffering, the decade of fallout for both major American political parties—was real enough, but the crash that sent everything sprawling downhill occurred entirely within the databases of financial institutions. None of it had to happen.

It started with real estate. President Franklin Roosevelt established the Federal Housing Administration (FHA) in 1934 with a mission to avoid the housing catastrophes of the Great Depression. The FHA insured lenders against default to limit their risk and established new borrower-friendly standards for mortgages nationwide. The thirty-year fixed mortgage rate, which became the new American standard, allowed families to keep up with their payments while pocketing decades of gradually swelling appreciation. No lender would agree to such borrower-friendly terms on the open market, but the U.S. government made thirty-year mortgages effectively risk-free for lenders. Real estate prices would appreciate forever, vouchsafed by the government and the financial system, and thus a worker who bought a house and paid off her mortgage and lived a stable, virtuous life could accrue some semblance of genuine wealth by the end of her working life. That wealth was not liquid, but it could be leveraged into debt that would allow the worker to send her children to college and finance a comfortable retirement—vital amenities that worker's wages could no longer cover. As infrastructure spending and other forms of direct investment dwindled, consumer debt was meant to step in and take up the slack.

As the housing market swelled, so too did the financial industry. Reagan's deregulating spree continued through the Clinton administration as policymakers chipped away at the venerable Glass-Steagall Act of 1933. Glass-Steagall erected iron walls around banks, separating commercial institutions (that provide banking services to regular people and businesses) and investment institutions (that seek only to extract profit for themselves and their clients) so that the economy would be insulated from the consequences of irresponsible investors losing their shirts on speculation. Thus the devastating financial crashes that kicked off the Great Depression could never be repeated. Decades of prosperity expunged legislators' memories of that awful era, such that, by 1999, the U.S. Congress repealed what little remained of Glass-Steagall and dissolved any meaningful distinction between commercial and investment banks.

President Clinton signed the law, figuring any policy that helped the banks had to be good for America. Republican senator Phil Gramm, who authored the bill, crowed, "Glass-Steagall, in the midst of the Great Depression, came at a time when the thinking was that the government was the answer. In this era of economic prosperity, we have decided that freedom is the answer."[1] The late Paul Wellstone, Democratic senator from Minnesota, lamented "Glass-Steagall was intended to protect our financial system by insulating commercial banking from other forms of risk. It was one of several stabilizers designed to prevent a similar tragedy from recurring. Now Congress is about to repeal that economic stabilizer without putting any comparable safeguard in its place."

It took less than a decade for global finance to destroy itself with greed. Before Glass-Steagall's repeal, investment banks like Goldman Sachs were sharklike, lean, and disciplined. They operated in their own environment, competing with other firms like themselves for profit opportunities. After the repeal, those firms found themselves thrown into a vast ocean with far bigger fish: the commercial banks, behemoths that had previously been kept out of investment markets and were now free to throw around huge sums of money. In order to compete, the investment firms needed to grow and grow fast, adding thousands of staffers at new offices across the globe, many of whom had little experience in their new markets and no cultural commitment to what ethics remained in the banking profession. Corruption flourished as the most successful bankers soaked up ever more of the system's rewards, trading assets they did not understand in patterns they hoped would make the numbers on their screens tick a little higher.

Many of those assets represented real estate; not the land or buildings themselves, but the debt associated with the real estate. American homeowners signed mortgage agreements with banks that then sold that debt—and the homeowner's promise to repay it—to other banks. Hundreds or even thousands of mortgages would get bundled into investment packages to be shuttled around bankers' computers as though they were actual objects holding actual value. To a trader behind a screen, what was the difference? It was all just numbers in columns, and the numbers were going up. Even as household incomes fell after 2002, real estate skyrocketed. From January 2000 to December 2006 home prices across America (as measured in a twenty-city composite by the Federal Reserve) doubled. Yet policymakers urged Americans to stay confident: the government had always protected the value of their homes before.

Housing prices ballooned, and families took on mortgages out of their price range because they had no other options, and lenders agreed they were good for it. Banks considered these "subprime" mortgages to be poor risks but issued the loans anyway assuming the government would backstop them. Many of those subprime mortgages had been packaged with other debts and sold as AAA-rated commodities, essentially risk-free, and now they constituted a significant portion of the financial positions of many banks—commercial and investment alike. The year 2007 saw an agonizing wave of foreclosures as people who had bought homes at the peak of the market fell behind on payments. In some states foreclosures tripled; nineteen out of every one hundred California homes received foreclosure filings.[2] When they tried to refinance their mortgages, they found that the banks that had once eagerly approved loans now nervously declared they could not assist. Millions of foreclosures proved to everyone that the real estate market was badly overvalued, and, just like that, the great real estate bubble popped.

Prices plummeted worldwide, leaving millions of people holding fixed mortgages they knew outstripped the value of their homes. Many simply walked away from their debts. Because there no longer existed any firewall between commercial and investment banks, *the entire global financial system* could no longer state with certainty what many of its assets were worth. Investment bank Bear Stearns was the first titan to fall in early 2008, and as Wall Street shuddered from that impact, it also became clear that Lehman Brothers, another investment bank that served as a major nexus for global finance, had been cooking its books. For years Lehman Brothers embellished its financial statements, claiming that the now worthless mortgage-backed securities it traded were ironclad.

The U.S. government made clear that Lehman Brothers would not receive a rescue package, so great was their malfeasance. The gruesome demise of one of Wall Street's leading institutions sent the rest into a spiral—after all, they were up to their necks in the same mud. Panic gripped the banks, the markets, the government. U.S. senator John McCain of Arizona, trailing badly in his presidential race against Senator Barack Obama of Illinois, announced he would suspend his campaign in order to focus on the unfolding crisis only to feebly backpedal when Obama declined to do the same.

Those of us without investment portfolios kept showing up to work, hoping to keep our jobs past the next payday. Anyone with an active mortgage could only hang on by their fingernails as the housing market shot up and then down again. A wholesale collapse of the financial system, the rest of the economy alongside it, seemed possible. Yet there had been no war, no natural disaster, or social calamity of sufficient magnitude to explain what happened. The invisible machinery undergirding our lives simply sputtered to a halt for reasons nobody seemed to understand, and nobody could fix. For many young people, this machinery became visible for the first time. Their outrage at the banks and the politicians who had recklessly enabled those banks helped propel Barack Obama to the White House, though the man was such a gifted communicator in retail and televised settings that he would have defeated almost anyone.

I was elated when he won, flush with the expectation that this inspirational figure would deliver on at least a few of his promises—chief among them, a proper national healthcare system. As much as the election of 2008 pivoted around the financial crash and bank bailouts that followed, the central running issue throughout the campaign had been a mutual acknowledgment by both major-party candidates that the American healthcare system was broken. It consumed far more resources per capita than any other advanced country's healthcare system while delivering far worse outcomes (still does, on both counts). John McCain sought to fix it through a combination of personal tax credits and further deregulation of the health market; Barack Obama promised a program of universal coverage through subsidies to employers and a National Health Insurance Exchange that would let anyone purchase thorough and affordable health insurance even if their employers did not offer it. It was not truly universal healthcare, but it seemed like a big and necessary improvement. To a nation scarred and made cynical by failed wars and government lies,

President Obama offered a credible alternative and a brighter vision of the future. I was twenty-four at the time, and it felt like there was so much work to do in this creaking country, so many important tasks left undone for so long, that the right leader could genuinely transform the way we lived.

Then What Happened Was

Obama started his presidential campaign as an insurgent challenger to Senator Hillary Clinton of New York, who commanded (alongside her husband, retired President Bill Clinton) more loyalty in the Democratic Party than anyone else. Knowing he could not rely on support from the party's elites, Obama built the largest grassroots fundraising and campaigning organization seen in decades: Obama For America (OFA). It won him the Democratic primary and helped win the general election as well, after which it boasted an e-mail list thirteen million strong with four million active donors. Not only had Obama's campaign won a relative landslide in the presidential race, the Democrats held their largest Congressional majority in a generation. "This would be the greatest political organization ever put together, if it works. . . . No one's ever had these kinds of resources," fretted former Reagan strategist Ed Rollins at the time.[3]

Yet once President Obama took his position in the White House, OFA went silent. For nearly two months they left their jazzed-up partisans flapping in the cold breeze. David Plouffe, the strategist who had run OFA through the campaign, wanted a break after the long campaign and decided to fold OFA into the existing Democratic National Committee. This might seem strategically sound—consolidating the Democrats' resources—but the Democratic Party apparatus held no affection for the grassroots that had so recently thrashed them in the 2008 primary. They weren't about to give up their positions and privileges to outsiders, and President Obama was too occupied with governing to fight the intramural skirmishes that would have been necessary to unseat those party elites. With Plouffe, his field marshal, heading for the exits, Obama's "army" sat directionless and frustrated, sniping at each other the way serious people do when leadership abandons them.

Obama struggled to implement his policies in the face of a brazenly cynical and racist opposition within Washington, yet he had functionally sent home the mass of organized people who had helped him prevail against establishment power. Perhaps Obama believed his best chances of success lay in working with the country's extant power structure, but in practice he turned his administration over to the same insiders whose intransigence and corruption had fueled Obama's own rise to power.

Obama's economic rescue package and healthcare plan both got mired in Congressional dysfunction, sabotaged by right-wing Democrats. When left-wing activists planned ad campaigns targeting the obstructing Democrats, Rahm Emanuel appeared in person to scream that they were "fucking retards" undermining the administration. Adept as Emanuel was at cussing, his record with legislation fell short of expectations. The Democrats passed a raft of halfway measures that were totally insufficient for the scale of economic crisis Americans were facing—the deepest, most prolonged crash since the Great Depression layered atop a rolling mortgage foreclosure crisis that began at the peak of the real estate market. Obama's policies made the crises even worse by bailing out banks while allowing them to foreclose freely on homeowners, often under fraudulent terms.[4]

Under the weight of many crises, having bargained away the most equitable solutions, was it any wonder Obama's Democrats got shelled in the 2010 midterm elections? They lost seven U.S. Senate seats, sixty-three U.S. Congress seats, six governorships, and enough state legislature seats to flip twenty state houses. Whatever window of opportunity President Obama ever had to change how this country was governed slammed shut forever. Rapid infusions of federal relief money might have kept many people in their homes or helped them recover jobs and wages lost in the financial crash, but Obama's impulses toward fiscal discipline pushed public investment lower than it had been even during the lean Bush years. From 2007 to 2009 non-military public investment held steady at 1.2 percent of gross domestic product (GDP); by 2013 Obama had cut that number in half.[5] At a time when borrowing costs had never been lower—spooked investors poured their resources into safe U.S. Treasury bonds, driving interest rates down to essentially zero. The largest, most obvious supplier of fresh investment capital stood idle. Millions of young people had their careers and future earnings permanently curtailed by job separations and long periods of forced unemployment. Eleven years and ten months would elapse before the bottom 50 percent of American earners recovered their precrisis income. Following the COVID crash of 2020, a stronger real shock to the global economy than occurred in 2008, more generous policies by Presidents Trump and Biden allowed that bottom 50 percent to recover in just twenty months.[6] We did not "lose" that decade; President Obama and his advisors chose to throw it away.

Rain When I Die

New York City is a space station. Everything consumed by its eight million residents is transported onto one of three islands accessible only by bridges, tunnels, boats, or aircraft. Business is transacted, but very little is actually produced on these islands; the economy is based on consumption subsidized by a wildly profitable financial industry that skims from actual production occurring elsewhere in the world. Most New Yorkers do not work in finance, but the city could not exist in its present form without this consistent source of outside capital. The island of Manhattan, a marshy little sandbar sandwiched between the Hudson River and a brackish tidal estuary called the East River, has been built up so extensively since the first

Dutch settlers in the seventeenth century that almost nothing remains of its original form.

Walking through Manhattan, one is always aware of the machinery under one's feet. Gratings hiss and steam; trains send tremors through the concrete at regular intervals. How deep would an excavator have to dig before encountering the island's natural muddy earth? Under the street lies a capillary lattice of ducts and pipes supplying water, electricity, steam, and fuel to nearby buildings. Under that layer is New York's transit infrastructure: not just subway tunnels and stations but ventilation ducts, water mains, and maintenance passages. Under those, 50–70 feet below the street, old brick-lined sewage mains are buried so their leakage doesn't seep into the water mains. Between those lines and the bedrock, one might find a layer of genuine Manhattan soil. The island's shores have been built out as much as a half-mile by heaping landfill into shallow water. Lower Manhattan's Water Street now lies two blocks from the shore.

Even Central Park is a wholly artificial structure built atop what used to be marshland. Every breath and heartbeat relies on a vast network of infrastructure, more so than in anywhere else in America. Whether a crisis goes well or poorly depends entirely on how well the government plans, builds, and manages. New York City is important to observe because it serves as a benchmark for American state capacity: if the wealthiest city in the nation can't handle a given problem, your local municipality probably can't either.

Hurricane Sandy rose from a low-pressure pocket of air in the southwestern Caribbean Sea during the third week of October 2012. On October 24 it became a hurricane over the sea south of Cuba and by sundown the following day it had killed sixty-seven people in Haiti, Cuba, and the Dominican Republic. Gulf Stream breezes carried Sandy up the east coast of the United States until a mass of high-pressure air over Greenland acted as a roadblock to jar it sharply west toward land. At that juncture it merged with a deep low-pressure system blowing in from the northwest to establish a "post-tropical cyclone" nine hundred miles in diameter—a hurricane sheathed in a winter storm.[7] Seasoned hurricane-chasing scientists in surveying aircraft gaped as their windows iced over; Virginia's governor issued a snow warning alongside a hurricane warning. Barometric pressures across the northeast dropped to record lows, the air pressure so altered that people began to complain of headaches and fatigue.[8] Superstorm Sandy, as media figures called it, represented one of the largest and most dangerous storms to ever strike the northeastern United States.

In many other countries, their national weather service coordinates with state broadcasters to define imminent threats and disseminate that information to the public. The U.S. National Weather Service has no such pipeline, instead relying on state and local agencies in addition to private media companies. In New York, differing interpretations of the forecast models left the state weather service uncertain of the threat Sandy posed until the storm made its deadly westward thrust. As New Jersey governor Chris Christie begged his constituents to flee, New York City mayor Michael Bloomberg declined to order evacuations. Storm experts looked on with alarm: half a million people live in New York City's highly localized flood plains.[9] "It was clear to me Bloomberg and the New York City Office of Emergency Management didn't understand the threat," hurricane specialist Bryan Norcross related to journalist Kathryn Miles. "There should have been warnings. . . . somewhere, there was a point of real confusion."[10] Officials issued "gale warnings" and "post-tropical depression" warnings, none of which adequately conveyed the danger Sandy presented to human life in such a dense urban environment. The region was about to suffer its worst storm surge in decades, and a full moon would cause high tides to rise even higher, but Mayor Bloomberg did not order evacuations until the day before Sandy made landfall. Even then, he ordered hospital patients and workers to stay put.

The year before, in 2011, Hurricane Irene had raked the eastern seaboard but diminished quickly, the flooding kept to a few low-lying areas. Seven people died in New Jersey, most of them located inland where rivers jumped their banks and limbs fell from trees.[11] Governor Christie ordered the Jersey Shore evacuated, and many people complied only to watch Irene spare their neighborhoods from any major damage. They accused the state government of upending their lives unnecessarily; a few had returned to find their homes robbed by opportunistic looters.

So when Christie issued another evacuation warning on October 26, 2012—three days before Sandy would make landfall—many Jersey Shore residents thought little of it. They cleared their yards, boarded up their windows, and hunkered down for more of the same. Some 70 percent of people who received evacuation orders ignored them, though many had no choice.[12] People who had nowhere else to go, or lacked the resources to secure transportation and shelter, or who lived with serious health problems will always be left behind in the absence of a proactive government.

The Jersey Shore suffered first: a chain of beach towns running the gamut from seedy to charming, none of them built for the storm now wading ashore like a cloud-wrapped giant. Winds whipped at ninety miles per hour, peeling the roofs off buildings as fourteen-foot waves tore sturdily built boardwalks to splinters. Every coastal region from New Jersey through New York City and up the length of Long Island felt an unprecedented storm surge. The U.S. Geological Survey recorded, "The peak water levels recorded at all stations in these areas also exceeded the FEMA one-hundred-year base flood elevations for these sites. In all cases, the peaks from Sandy have surpassed anything documented previously at these sites."[13] As often happens with storm surges, the water arrived with almost supernatural speed. In minutes a clean street could become a waist-deep torrent sweeping up every untethered object and creature. Water trapped people in their homes; it snatched vehicles from the street; it tore children from their parents' arms as the hurricane winds droned unbearably from every direction.

Just to the north, New York City was taking on its own water. The city has flood protection measures designed to withstand eight to ten feet of storm surge, and most public officials considered the risk of a higher surge very unlikely. Yet should those measures fail, should waves overtop the walls, the consequences for the city are always severe. As discussed earlier, nearly all of New York's infrastructure—not just water and sewage but transit and heat and electrical—lies underground. Tunnels all the way down, and when water gets into this aging space station it does a great deal of damage. New York's antiquated subway system, where most of the flood water drains first, flooded in 2004 and 2007. Sandy swamped it completely: "the worst disaster to hit the subway system in its one-hundred-eight-year history," declared transportation authority official Joseph Lhota. Fourteen feet of water hit lower Manhattan.[14]

People drowned in their homes. They drowned in their cars, in their garages, in storm drains where flash floods swept them screaming away from anyone who tried to help. Downed powerlines electrocuted some who tried to get help and kept rescuers from reaching their burning bodies. Hospitals lost power, and staffers had to carry patients up darkened stairwells so they would not drown in their beds. More than two million people lost electricity for days, if not weeks. The subway system still labors from the damage Sandy caused, as do many places in New Jersey, Staten Island, and Long Island where locals don't have the capital to rebuild.

Americans in the afflicted states woke to scenes of total destruction typically associated with South Asia, or the Caribbean, or perhaps New Orleans. Many people had no personal experience with damaging hurricanes; many found their flood insurance paid out only a fraction of the damages. Hurricane Irene should have served as a warning the year before, but, because it fell one foot short of overtopping the city's storm barriers, most dismissed the threat.[15] Millions of people could no longer ignore the fact that they were only as safe as the local infrastructure allowed them to be. That infrastructure, as Sandy demonstrated, suffered badly from decades of underinvestment. In the face of a climate change–driven, twenty-first-century storm, it failed to the tune of $65 billion in damage and more than 150 people killed in the United States alone.[16] Over a year after Sandy subsided, thirty thousand of the people it rendered unhoused remained so. Federal aid programs spent a great deal of money, but much of it was devoured by insurance companies and red tape. No single homeowner received more than $36,000 from FEMA's emergency aid fund, as federal officials cited post-Katrina regulations intended to thwart fraud.[17]

Just as the federal government learned little from Hurricane Katrina, it refused to adapt following Hurricane Sandy. The U.S. Department of Housing and Urban Development and Army Corps of Engineers invested hundreds of millions of dollars into higher flood walls for New York City. Some low-lying areas destroyed by Sandy got bought out and converted into coastal parks designed explicitly to mitigate storm surge. The Army Corps has proposed more exotic proposals like a six-mile sea wall protecting New York Harbor or mechanized gates sealing off gaps between Long Island's sand barrier islands, but every expert agrees that these expensive interventions will accomplish little in the face of a persistently rising sea.[18]

The year 2021 saw Hurricane Ida visit the New York metroplex in a visual (if not meteorological) recreation of Hurricane Sandy. Ida took a different path, shredding towns along the Louisiana and Mississippi Gulf coasts in the last days of August 2021. The Corps of Engineers' flood protection system held up in New Orleans, but anything lying outside their battle lines suffered severe damage. Ida maintained its strength over land, traveling northeast, depositing record rainfall as it crossed the eastern United States. Hurricane Henri had soaked the Northeast just a week prior, and Hurricane Fred passed through the week before Henri. Soil was inundated, streams ran high. The pairing of Henri and Ida recalled the awful Delaware River flood of 1955 (see chapter 6), where one storm swelled the river and the second drove it mad.

Sandy hit New York City from the sea, overwhelming its coastal barriers, but Ida struck from inside them or, more accurately, from above them. New York caught three inches of rain in a single hour—the wettest hour of the wettest day in its recorded history. Over seven inches would fall in sum, more than the city's drainage system was ever meant to take so quickly.[19] As occurred with Sandy, the excess water rushed down through New York's layered strata of machinery to reach the wide open spaces of its transit system. What measures had been completed after Sandy were aimed at controlling a coastal storm surge, not inundation of the drainage system, and city officials acknowledged any rainfall exceeding two inches per hour would overwhelm their safeguards.[20]

American social media shuddered at phone-captured videos of subway tunnels rapidly filling with a mix of water and sewage. Dozens of New Yorkers suffered horrific drowning deaths in basement apartments where persistently rising rents had driven so many people. Entire families died screaming for help while their neighbors tried to fight through standing walls of water. Ida killed fewer people than Sandy, yet the economic damage to the United States was $65 billion—almost exactly the same.[21]

For Lack of a Marsh

Even allowing for the ever-marching catastrophe of anthropogenic climate change, New York City's vital infrastructure should be in much better shape. Yet the "space station" approach its planners have pursued has destroyed the natural barriers that should safeguard the city from rain and storm surge.

We do not think much about wetlands, not unless we are presently thigh-deep in one. In the United States, a country built on the maintenance and manipulation of real estate prices, wetlands appear as negative spaces on maps: darkened plots that produce nothing and have no value. The expensive process of reclamation might dry them, level them, and render them suitable for investment, but when the rain falls hard the water tends to pool in the same low places it always did. Rare is the person who will get excited about marshland. It's inconvenient to navigate around, smells bad, isn't colorful or obviously beautiful like so many natural environments. It doesn't not make a great argument on its own behalf.

Yet marshes hold value no appraiser will ever record. Through them swim countless: fish and frogs and shrimp and diatoms and bacteria. Decomposition's constant churn gives micro-organisms all the food they want. Plants grow densely in shallow water, even if it is brackish, and flying insects play the air like a rosined string. Birds hunt those bugs and shelter in the vegetation alongside rabbits and other small mammals.

Wetlands boast some of the greatest density of plants, animals, and microbes found in any ecosystem. But wetlands also have real, quantifiable value that any resident of our rapidly warming planet should appreciate: they are ideal flood barriers. When heavy rains or storm surges cause a sudden influx of water into a region, it pools at low elevation. If the ground at those low points is impermeable to water (if it is paved, for example) then the water accumulates rapidly and seeks to flow elsewhere. Given sufficient volume, that flow might be a destructive flood. If, on the other hand, the flood plain is left in its wetland state, it absorbs new water like a sponge. Porous soil fills until saturated, held in place by a dense picket of vegetation that drinks and stores water on its own. Even if water overflows the marsh, it will move more slowly and at lesser volume. Every acre of marshland paved over makes the surrounding area more vulnerable to flood. According to the Environmental Protection Agency, wetlands adjoining the Mississippi River that once accommodated sixty days of flood water now hold only twelve days' worth due to riverbank development.[22]

New York City used to be surrounded by some of the biggest, densest, most productive coastal wetlands on the east coast. Developers filled in almost all of it during the nineteenth and twentieth centuries. Flat dry land was so much more valuable than marshes, they'd have been crazy not to. Eastern Brooklyn and southern Queens were, for most of the city's history, the outermost reaches of a brackish wonderland known today as Jamaica Bay. European-descended settlers lived by fishing and working in industries that benefited from proximity to the growing city but could not be sited within it. Fertilizer and fish oil factories that made the surrounding air smell intolerable sprung up on undeveloped sandbars with names like Barren Island and Dead Horse Bay. Hundreds of personal and commercial watercraft trawled the deeper channels for fish at high tide.[23]

By the twentieth century, dense development pushed up to the waterfront. Islands had been joined to the coast by means of landfill bridges, coastal marshes got filled with earth, and the island-spotted sea shrunk to a more recognizable bay. A total of 85 percent of New York City's salt marshes and 99 percent of its freshwater marshes presently lie buried under asphalt.[24] Erosion and rising seas claim another six acres per year. To alleviate congestion at New York's LaGuardia Airport, the city began building a second major airport at Idlewild Beach, on the shore of Jamaica Bay. Idlewild Airport opened in 1948, converting several square miles of Jamaica Bay into a landmass seemingly summoned from the water itself, protected by high levees. By the 1960s it was the second-busiest airport in the United States.[25] President Kennedy's assassination inspired a name change to John F. Kennedy International in December 1963.

In 2012 Hurricane Sandy sent Jamaica Bay over the levees and onto the tarmacs, forcing a wholesale closure of JFK alongside every other airport in the region. Long Island Sound lurched over LaGuardia's barriers, flooding the older airport so badly it had to close for several days and reopen at reduced capacity. The federal government maintains a list of the U.S. airports most vulnerable to coastal flooding; all three of the region's major airports (JFK, LaGuardia, and Newark, New Jersey) sit on that list.[26] Given the known vulnerabilities of New York's bridges and underwater tunnels during severe weather, it's easy to imagine a storm bad enough to effectively halt the transit of goods and people in and out of the city. Hundreds of thousands could potentially be trapped inside an unfolding humanitarian crisis.

New York continues to build higher. It is a process that began more than a century ago and will continue so long as humans inhabit the region. For most of the twentieth century, it seemed the city had won its race with nature, but the ocean is nothing if not persistent, and the technology keeping the metropolis afloat is failing. If New Yorkers thought their home had grown more resilient since Hurricane Sandy, Hurricane Ida laid those notions to rest. A sufficiently forceful storm will overwhelm the city's systems; this has been known and acknowledged by oceanographers and engineers alike since the early 2000s.[27] Its leaders know this and simply hope to make it through the next hurricane season without calamity. They install hydraulic-powered flood gates; they buy up shoreline properties and convert them to "parks" that are actually pleasant greenery laid atop sophisticated modern flood walls that inhibit the movement of water both above and below the ground.[28]

These solutions would have mitigated the damage from Sandy or Ida, but, in the age of warming and rising seas, a bigger storm is always coming. Even if global climate change were halted expeditiously, New York City would still face decades of worsening conditions and require infrastructure investments into the hundreds of billions simply to remain inhabitable for all its residents. The city needs to restore its wetlands in a hurry. Yet expanding the coastal park network and optimizing other parks for flood control is not really on the table, not at scale. No industries will be made to vacate their riverside lots, and nobody is going to rip up the pathways of Central Park to restore it as Manhattan's wet marshy core. New York spends 0.6 percent of its annual budget on parks when less flood-vulnerable big cities spend up to 4 percent.[29] The city must remain a space station because nobody who wields power wishes to see it fundamentally change, and nothing short of that change will suffice to protect it.

Like so many institutions in the twenty-first century, the city's power centers prefer slow, grinding collapse to any of the realistic alternatives. As Alyssa Battistoni wrote in her excellent post-Sandy essay, "The Flood Next Time": "Disaster is supposed to demonstrate what the science tells us we can expect from a warmer world. But the problem with disaster as education is not that everyone learns the same thing."[30]

Realm of Water

Almost every city on the east coast stands to suffer in some way from rising seas. All of them are vulnerable to flooding, whether the water comes from the sea (storm surge) or the sky (sudden deluge). Given the outsized attention paid by the press to ocean level rise relative to other indicators of climate change, the typical American could be forgiven for thinking of floods as a coastal issue. Yet the entire eastern half of the country is at risk. The Mississippi River basin and everything east of it constitutes the "wet" portion of America. That half boasts most of the surface water and receives most of the precipitation each year.

Climate change doesn't just make temperatures warmer on average; it makes existing weather gradually more extreme. Storms get wetter while droughts grow drier. As a consequence, the flood control measures we currently rely on will no longer sufficiently protect us. The many thousands of dams and levees erected during the last hundred years will not necessarily collapse, but the water will overtop them. Home-destroying, death-dealing floods—monsters once mostly banished from American life by virtue of the planet's largest inventory of dams—will return, and a society that has forgotten how to build or even plan for the future will suffer as a consequence. The eastern half of the United States is facing an era of flooding the likes of which few living people can recall. The damage will be unpredictable and uneven in their distribution, yet we can be sure of its severity. Those Americans who live in wet regions will find their lives growing more difficult as time goes on. Heavy rains may complicate your commute or wash your home entirely off its foundation. Far from the coast, in the hills of West Virginia and Kentucky, archaic infrastructure spews sewage into local rivers with every major storm. The rain seems to start earlier each year and end later. Federal resources are hard to come by, especially since FEMA flood maps tend to ignore the mercurial streams and creeks that course by the thousands through Appalachia. More than a century of deforestation and strip mining has depleted the forests that might have slowed the floods.[31]

In these towns, as in New York City, local politicians and citizens resist structural changes on the grounds that altering too much too quickly will undermine the industries on which each place relies: coal mining in Appalachia, banking in New York. The implication is that no other industry can replace them, and the alternative is destitution. This is emphatically not the case. It is an argument that ignores both the tremendous (even overwhelming) tally of infrastructure work to be done across this country and the value in that work. The coming century does not have to be filled with tragedy, with long-foreseen disasters and heartbreaking death tolls, with the sense of inundation that so many of us feel every day. We could always have stopped the floods, and we still can. The costs of avoiding this work vastly exceed the costs of simply paying Americans good wages to do it.

10

DESICCATION

Rain pocked the dust last night at my home outside Sacramento, more than eight months since the last time it visited. But for one scruffy thatch of crabgrass the yard is bone-yellow, and it disintegrates at a touch. The soil is so hard that that shovels cannot bite; since the winter it has acquired ridges, dips, and rises. Drought has revealed the mean, gnarled creature hidden under prosperity's coat. For twenty years now drought has gripped the American West. From the Rockies and western Texas to the Pacific, nearly half of the United States struggles to secure enough water while the other half sweats out flash flood warnings each hurricane season.

The West has always been dry. While the eastern portion of this country enjoys precipitation year-round, any given inland area in the western United States typically goes for months every summer with very little rainfall. Winter is the key; snow falls hard on high mountain ranges and melts during the spring, swelling great rivers like the Columbia and the Colorado that distribute that water across tens of thousands of square miles. Dry winters quickly tip the entire ecosystem out of balance. Diminished snowpack in the mountains causes faster melting and evaporation, since the snow reflects the sun's heat whereas stone absorbs it. Drier, hotter soil absorbs runoff before it can reach river beds. The rivers themselves become slow, shallow, and warm, encouraging algae blooms that devour oxygen and can poison wildlife. If global climate patterns persist in such a way that the West does not receive sufficient winter precipitation for several decades, the entire ecosystem faces collapse.

These "megadroughts" have happened before: four times in the last twelve hundred years, including one drought centered in northern Mexico that lasted for more than a century. Droughts ruin crops; megadroughts may sink civilizations. Thriving Indigenous populations like the Ancestral Puebloans in the Southwest had their societies brought low by megadroughts. In the face of relentless deprivation, their only choice was migrating out of the region, abandoning their impressive stone structures for different lifeways elsewhere. Some depopulated areas would not be inhabited again for centuries. Our present water emergency is unique in that it represents a cyclical process combined with carbon-driven warming. This drought afflicts all the land west of the Rockies, but it is most concentrated in the Colorado River basin. That basin, as covered extensively in this volume, constitutes the most economically valuable water system in the western United States. Crucially, the twentieth century (the precipitation records of which form the basis of the 1922 Colorado River Compact's apportionment figures) stands out as the region's wettest period in more than one thousand years. When drought struck, something had to give.[1]

Which is why in August 2021 the U.S. Bureau of Reclamation declared the Colorado River to be in a shortage for the very first time in its history. That announcement had no immediate impact for most Americans beyond some farmers in Arizona who saw their Reclamation allotments cut, but to observers of western water policy it sounded a grim knell.[2] The Compact's time had reached a bitter end more than twenty years into the drought. The conditions it describes no longer reflect reality, and nobody expects that to change anytime soon. As with so many other American institutions, the agreement as written can no longer be kept, but neither can it be legally discarded. In that gulf between legal fiction and hard reality stand many millions of Americans just trying to navigate their lives.

Bathtub Rings

The rings are the hook to every news story about the Colorado's declining reservoirs; the rings are the first thing you see. Walls of white mineral deposits stand above the waterlines of Lake Mead and Lake Powell, the largest and second-largest reservoirs in the United States as though painted with a gigantic brush.

That "bathtub ring" appears any time a rocky reservoir drops below its peak. As droughts persist into megadroughts the ring becomes an unnerving visual reminder of how much water is missing. At this writing, near the end of 2021, the ring at Lake Mead stands 160 feet high. At Lake Powell it is 155 feet. Human beings boating past the walls are absolutely dwarfed by a powerful and unmistakable omen of doom. Obviously, no one is happy about it. Lake Mead is barely one-third filled and Lake Powell has ebbed below 30 percent of full pool.[3] As Lake Mead drops to levels not seen since it filled, the drought has exposed immense intake pipes and brought gruesome artifacts to light: decades-old mob hits and drowning victims, the hulks of abandoned pleasure craft.[4]

It wasn't supposed to be possible. If the era of dambuilding out West had a single objective, it was to armor the seven Colorado River states against drought: build up reserves during wet years, draw them down during dry years, and, over decades, nature should work everything out. But federal planners never adequately planned for a twenty-year drought and were hamstrung all along by the Colorado River Compact's mistaken assumptions. The agreement assumed a river providing 16.5 million acre-feet annually, which it managed to do with some consistency in the twentieth century but rarely before and never since. The Compact's structure also incentivized states to consume the entirety of their legal allotments lest other states take the water for themselves. The end result is a network of reservoirs drained to the barest threshold of functionality. Mead and Powell have never been lower than they are now, not since they were initially filled following the construction of the Hoover and Glen Canyon Dams.

One of the Bureau of Reclamation's cleverest maneuvers during the age of big dams was linking electricity to water—the two most important resources for economic expansion during the twentieth century, secured and delivered by means of high dams that stored the water and used it to drive electric turbines. It let the bureau pretend their unprecedented infrastructure binge came at no cost to the taxpayer, and it secured the bureau's position in the political economy of the American West. Yet the age of drought has revealed that marrying those resources in a union of concrete imperils them both.

Every modern dam includes outlets for the reservoir behind it, so engineers can release water at will. Even dams that do not produce electricity need this basic feature to safely operate. In the case of a hydropower dam, these openings sit well below the dam's crest but also above where the dam's base meets the reservoir floor, so sediment from the floor doesn't get drawn into the turbines. That means that every reservoir has a threshold below which water cannot move past the dam: the "dead pool" level.

If water cannot move past a reservoir, the river is no longer a river. Nothing can be distributed past that point until the reservoir fills past dead pool again. Engineers can postpone that day by releasing water from dams on tributaries upriver, but those reservoirs are small compared to Lakes Mead and Powell. The federal government's resources may break before the drought does. In the fall of 2021 a series of monsoons in the Colorado basin washed over a million tons of sediment into the river. Typically the Bureau of Reclamation responds to similar conditions by releasing a pulse of water from Lake Powell, sweeping the sediment downstream to Grand Canyon National Park where it shores up constantly eroding riverbanks. Yet in this case the reservoir stood so low that federal authorities canceled the water release, stranding at least a million tons of sand on Powell's floor.[5] When reservoirs approach dead pool, every system around them begins to fail. Two of the most important powerplants in the Southwest, supplying more than ten million Americans, are two or three more bad winters from going dark.[6]

Above California's Central Valley, among the oak-shrouded Sierra Nevada foothills north of Sacramento, the mighty Oroville Dam distributes power throughout the state. Since 1967 it has stopped the Feather River behind a 770-foot dam: the tallest in the United States and thus one of the very best when it comes to hydropower. Below the dam, in a cave that could accommodate the state capitol building, six mammoth turbines turn falling water into electricity. But the twenty-first century megadrought has steadily depleted Lake Oroville, left its trees standing abashed above a naked hundred-foot "bathtub ring" of dry clay. At the bottom runs a thin ribbon of clear water—the Feather River, refusing to die. The lake has dropped too low, and, in August 2021, the Oroville Dam reached inactive pool for the first time in its history.

Yet four-and-a-half years prior, in 2017, an exceptionally wet winter nearly destroyed the Oroville Dam. Climate change has warmed the atmosphere in the West to the point where weather patterns become increasingly unpredictable, so even amid a megadrought certain areas may receive historically intense rain. The winter of 2016–17 dumped a heavy snowpack onto the Sierra Nevada mountains, followed by a warm spring that sent most of that water rushing into California's reservoirs. Anderson Dam, located above the city of San Jose, overtopped and flooded several neighborhoods. At the Oroville Dam, engineers sent so much water down the dam's seldom-used spillway that the concrete structure cracked in half. Tons of stone and debris crashed into the Feather River, imperiling millions of salmon in a state hatchery just four miles downstream.[7] Evacuation orders went out to tens of thousands of people, many of whom learned for the first time that they lived in a vulnerable floodplain. Thankfully the worst did not come to pass; workers reopened the damaged spillway, and the skies declined to test the dam with further rain. The fish hatchlings were safely evacuated, and nobody ended up homeless, but, if not for lucky weather, the tallest dam in America might have fallen.

When the Desert Wins

Decades of ground seep, evaporation, and unsustainable agricultural practices took their tolls on Lakes Mead and Powell even before the last twenty years of drought. From 2000 to 2020 Lake Powell's water level sank by 140 feet. Houseboats are barred from certain sections of the lake which have become too shallow. Several boat ramps have closed, and, on some of the Colorado's smaller tributaries, even kayaking has become impossible. Hazards multiply as the submerged natural world slowly comes back into view. For decades the wooded glens for which John Wesley Powell named Glen Canyon lay entombed under hundreds of feet of water. Leaves and stems rotted quickly but the trees remained, unable to decompose in the low-oxygen environment the reservoir established. Today those long-dead trees reach toward ever brighter sunlight. They lurk just below the surface, threatening watercraft like hidden reefs.

Barring a miraculous change in the West's climate, it seems likely Lake Powell will cease to exist during this century. The question is not how to save Lake Powell—no one has that power now—but rather what shape the future takes. Unthinkable as it would have been twenty years ago, a growing chorus of activists and scholars has been urging the federal government to remove the Glen Canyon Dam. Their plans would consolidate the Colorado River's storage capacity in Lake Mead, as a sort of strategic retreat from the megadrought.

Let's consider what happens if nothing substantial changes about the Glen Canyon Dam. First the boat ramps will be left high and dry, as Floyd Dominy's beloved recreation disappears along with the tourist money it lured from families in Las Vegas, Pasadena, and the Salt Lake suburbs. Next the hydropower will cease to function, and, finally, the Glen Canyon Dam will lose its capacity to release water downstream. The water, currently quite cold, will become progressively warmer. It will also leech toxins from the decades' worth of silt piled at the bottom of Lake Powell: millions of tons, some of it toxic, choking the flow of a river too weak to heal itself. Nobody will ever pay to clean it, so that silt will clog the hallowed passageways of Glen Canyon for the rest of human history. The fish and birds living at the lake will die or depart, leaving only rafts of stinking algae on bathwater-warm shoals.

Facing shortages of water and electricity, civilization in the Southwest can only retreat. One need not imagine a *Mad Max* landscape of perpetual conflict to appreciate what deprivation means in places America has written off and chosen to forget— cities like Flint, Michigan. High prices for power and water will squeeze people already struggling with lacking job opportunities. Agriculture, power generation, and water management are crucial pillars of these communities. In their absence, entire towns will dry up and gradually blow away.

Some of the nation's most productive agricultural centers, like California's Imperial Valley, will face crises of profitability as they pay more and more to secure water the Colorado River system can no longer reliably supply. Nearby is the Salton Sea, a shallow depression near the Imperial Valley that historically served as a flood basin for the Colorado River and filled with water after the Imperial Valley's first irrigation canals ruptured in 1905. Since then it has gradually shrunk, replenished mostly by runoff from local irrigators, and, if the Colorado system collapses, the Salton Sea will evaporate for good. Today the lake is a valuable habitat for migratory birds. allowing it to vanish would not only harm those birds but also expose a noxious hazard: bottom sediment tainted by pesticides after decades of agricultural runoff. In the Central Valley's sun and wind, that sediment will rise as fine particles and lodge in the lungs of everyone in the county (which already has some of California's worst asthma rates). California's government acknowledges Salton's dire need for restoration but cannot stomach the $9 billion price tag.[8] It seems unlikely that the drought's chokehold will open the state's purse.

Megadroughts are always painful, but avenues do exist to real change, and a more prosperous tomorrow for everyone in the Southwest. Perhaps the most promising, and the most popular, also happens to be the simplest: remove Glen Canyon Dam, or open its gates such that the river flows unobstructed. Retiring the second-largest reservoir in the United States might seem like a drastic step. Opening the Glen Canyon Dam, to say nothing of removing it entirely, would represent a fundamental challenge to the Southwest's economic order. It would be extremely difficult to convince anyone in Page, Arizona, that their city should suffer and diminish for the sake of running a massive interstate system more efficiently. Imagine telling the legislators of the upper basin states that they should give up their water security (even if it is not really secure) in order to promise more water to the lower basin states. But the megadrought facing the American West demands urgent responses, and, in truth, the Colorado system has little to lose. Defenders of the status quo argue not so much with environmentalist "radicals" as with the drought itself.

Lake Powell will drain, sending a great but gradual tide down through the Grand Canyon until it reaches Lake Mead. Mead—as perfect a dam site as God ever made, situated in a narrow "gunsight" canyon of impermeable rock, adjacent to Las Vegas—will remain untouched, though of course it would fill higher than it has in decades. Boating and tourism will diminish at Lake Powell, but that scenario is already upon us as the megadrought strands entire fleets of gleaming white house-boats and bars even kayaks from Powell's tributary canyons. The region would need a new source of electricity, but the Hoover Dam's increased head would make up a good chunk of Glen Canyon Dam's lost production, and moving electricity long distances is far cheaper than doing the same for water. In any case, the water Lake Powell loses to evaporation each year is more valuable on pure dollar terms than all the electricity the dam produces (see chapter 5).

One must not be too sanguine about what an undammed Glen Canyon will actually look like. Under Lake Powell's crystalline water lies a badly polluted canyon smothered in silt. Millions of tons of sediment that should have been feeding marine life and shoring up river banks all the way to the Colorado's denuded delta on the Sea of Cortez instead languish behind seven hundred feet of poured concrete. Nearly sixty years of deposits sit on Lake Powell's floor. The region is rich in heavy metals, and, instead of being harmlessly dispersed, they have been allowed to concentrate. Abandoned uranium mines lie submerged. Hydrologists estimate the silt load could be cleared in as little as a decade by allowing the Colorado to run wild as in the past, but what a brutal decade that would be: tides of muck shining brown and red in the sun, stinking to high heaven, much of it requiring professional cleanup to avoid poisoning fish and plants downstream.[9] Only once the mess cleared could anything approaching the original ecosystem begin to appear. One of the nation's favorite vacation spots would look bad and smell worse. Even decades in the future a revived Glen Canyon would offer a fraction of the tourist amenities Lake Powell does now.

Removing the Glen Canyon Dam to fill Lake Mead would be an historic victory for the conservation movement, but those who support the plan on its merits must reckon honestly with the principle of democracy. Even if one believes that the Bureau of Reclamation has subsidized an unsustainable level of development in the West, those states' "excess" population now numbers millions of people who will not simply accept efforts to remove the infrastructure on which they rely. The water Lake Powell provides to northern Arizona cannot just be piped uphill from Lake Mead the way electricity can. Whatever direction we choose must be democratically valid to succeed in the long term, even if policies of the past—like flooding Indigenous lands to establish many of the nation's reservoirs—were imposed by coercion. That inherent tension infuses so many issues of conservation and planning for the future; if a balanced resolution were easy to find, we'd be living in a very different world.

Fortunately, urban centers have proven to be better conservators of water than industrial agriculture. Per capita water consumption has fallen by more than 20 percent in arid cities like Phoenix, Las Vegas, and San Diego, while agriculture continues to use more than 70 percent of the water in those states.[10] Technology helps urban populations reduce their water use, yet it has consistently failed to impact agricultural consumption. In 2021 a team of agricultural economists reviewed two hundred papers from around the world. They found that while water use per hectare decreased with the use of technology like drip irrigation, farmers responded by placing more land into cultivation. Growers used just as much water as before, and in some cases even more.[11] When they ran out of surface water, they dug deep wells to mine irreplaceable groundwater reserves. The researchers declared the notion of achieving real water savings through technology a "zombie idea," which refused to die no matter how often it was disproven.[12] Water savings will only come when the government takes an active role in economic planning. Modest water supplies can support surprisingly large urban populations, and the moral case for prioritizing human consumption is easy to argue. The question really becomes which industries the federal and state governments choose to subsidize.

Do we really need to practice European-style agriculture in the country's most inhospitable places? Arizona uses most of its agricultural water to raise cattle, lettuce, cotton, hay, and alfalfa. New Mexico's largest crops are hay, alfalfa, and wheat. Many of these crops—lettuce, cotton, and alfalfa in particular—consume a huge volume of water relative to their market prices. Their growers stay in business because the federal government footed the bill to develop their water resources and maintains a price floor by paying growers in the water-rich eastern United States to reduce their production. In an economic environment where goods can be transported cheaper and more efficiently than ever before, why does the Southwest need to grow its own alfalfa? To cheaply feed their cattle, one might answer, but why are cattle grazing these arid states to begin with? Perhaps in the twentieth century, flush with historic precipitation, it was a luxury we could afford. But how, in the worst megadrought of the last one thousand years, could cows be more important than human beings? Why are private interests allowed to draw freely from underground aquifers that will not recover for centuries? And why are companies able to obtain cheap water for semiconductor plants and data centers (both of which use large volumes of water for cooling) in the Arizona desert? Choices must be made, and ordinary people have suffered enough in the economic planning of the last fifty years. Government policy created these incentives, and only pursuing a different policy can change them. It is time to root our imaginations in the world we actually inhabit.

Lake Powell does not generate any extra water—if anything, evaporation and ground seep make it a losing proposition. It stores water as a buffer against drought, but that buffer is already exhausted. If the Colorado River's inflows are truly insufficient to sustain the human population of Arizona without Lake Powell's storage and distribution, then the region will decline no matter what the state does. Emptying tributary reservoirs to keep Powell on life support leaves more remote communities without any water source; this ought to be avoided. New industries must rise to replace the old ones. The seven states that signed the 1922 treaty governing the river receive more sunlight than almost anywhere else in the continental United States at a time when solar infrastructure has never been cheaper to build, and government borrowing costs are historically low. Arizona should be a major exporter of electricity regardless of whether the Glen Canyon Dam operates. The same principle applies to the other six states. An enormous federally funded build-out of solar capacity would go a long way toward soothing the economic anxieties of Americans who worry change will leave them behind.

The Bureau of Reclamation gestures sternly toward charts and numbers hoping consumption falls, hoping the coming winters will fill their reservoirs and help the old ways make sense once again. But hope is a prison: an admission that we are helpless in the face of the future. The Colorado River system as it stands will die in agony and establish a new Rust Belt in the Southwest as the unmitigated fallout ruins people's lives. By the year 2100 climate change could cost the U.S. economy $2 trillion per year in lost productivity, with the bulk of that toll falling on the West and Southwest.[13] The old ways will end; nothing can stop that now. Let us unshackle the river while it still has strength. Let us resurrect Glen Canyon and build a solar-energy empire. We have just a few years left to fix some of our mistakes and restore what was lost, at a cost no greater than what inaction will already take. Nothing could be more radical or more dangerous than allowing the status quo to persist. That is the choice we face in the drought-stricken West: a future partnered with the desert, or none at all.

The Fish that Holds the World

"My strength is from the fish," Chief Weninock of the Yakama Nation said of the animal we call salmon. "My blood is from the fish, from the roots and berries. The fish and game are the essence of my life. I was not brought from a foreign country and did not come here. I was put here by the Creator."[14] The Indigenous people of the Pacific Northwest have long understood the primacy of salmon in their part of the world. When the earliest salmon of the season reached their nets, they would take a single large specimen—the "first salmon"—and make it the centerpiece of a festival lasting up to ten days, during which further fishing was forbidden.

Antone Minthorn, a leader of the Umatilla Nation in Oregon, explains, "The importance of the first salmon ceremony has to do with the celebration of life, of the salmon as subsistence . . . it is the natural law; the cycle of life. If there was no water, there would be no salmon, no food, there would be no cycle."[15] The salmon's cyclical life holds special symbolic resonance. Spawning in freshwater streams fed by cold mountain runoff, salmon fry spend up to six months in their natal waters before making their way down tributaries to large rivers. The baby fry become juvenile parr and then adolescent smolt, feeding and growing as they travel. More than fifty million years of natural selection have made them athletic and voracious predators. By the time they reach the sea, they are ready to feed on larger prey, and their bodies have adapted to survive in salt water.

Up to five years in the open ocean swells them like summer hogs. Once fully mature and ready to breed, the salmon use their sense of smell to retrace their formative path—up the big rivers, up the tributaries, up the streams, leaping over rocky barriers, climbing ever higher alongside legions of their cousins. Nothing in these streams can feed them anymore, but they're done eating. Frenzy grips their brains and bodies, a burning need that drives them onward. Fat reserves, built up in the sea, will carry them as far as they need to go. Up in the hills where they first spawned—as much as two thousand winding miles from salt water—they mate and deposit millions of eggs in gravel nests called redds. These glowing amber orbs make tempting meals for freshwater scavengers, but sheer numbers will ensure their survival until spawning season. Totally spent and exhausted, their final task discharged, the salmon die. Their bodies, choking the streams, seed the waterways with nutrients their offspring will recoup on the path back to sea.

Seven species of salmon exist, six of which inhabit the Pacific coast of North America (the Atlantic salmon is the seventh). Smaller species like the pink salmon and Masu salmon rarely exceed two feet in length; the mighty Chinook can grow to five feet and weigh well over one hundred pounds. Salmon spawned during a given breeding season will return to breed during that same season years later, so the fish sort themselves into spring, summer, fall, and even winter "runs" that repeat over time. Winter run salmon spawn more winter run salmon, summer run salmon spawn more summer run salmon, and so on. Since migrations take place over several months, late salmon from one season's run have been known to mingle and occasionally interbreed with early salmon from the next run, but scientists see these populations as effectively distinct. Indigenous Americans believed the same. To their understanding, each season's run represented a different family of salmon—a family in the literal human sense, since they recognized no spiritual distinction between human and animal lives. A salmon was not a fish but a bearer of gifts, a friend using his life to sustain another. The "first salmon" ceremony honored that sacrifice and celebrated the centrality of the salmon's lifecycle to that of human beings. Its prohibition on fishing during the festivities allowed a portion of each season's run to pass unharmed, ensuring the salmon family's continued health.

Once fishing began in earnest, the tribes harvested salmon with near-industrial efficiency. Anthropologists have concluded that, prior to colonization, the salmon-fishing nations of the Pacific Northwest consumed one thousand pounds of salmon per capita every year—half a ton averaged across every man, woman, and child. When Chief Weninock said, "my blood is from the fish," he did not speak metaphorically. Children learned to fish with dip nets and gill nets while their elders worked on more elaborate schemes: rings of stones arranged in shallow riverbeds to slow and expose salmon, wooden weirs laid bank-to-bank in deeper currents to gather and herd them, elaborate "fishing wheel" contraptions that hoisted salmon from the water and dropped them into baskets. A population of just twenty thousand people could eat four million fish annually, and that didn't even include fish harvested for trade.[16]

Tribal elders coordinated fishing operations for their communities, telling family units when and where they could fish, accounting for the size and quality of that season's run and other factors like how much space stood empty on the tribe's drying racks. The whole operation was meticulous, highly productive, and sustainable over thousands of years. Ecologically speaking, the salmon lifecycle represents a constant transfer of nutrients from the Pacific Ocean to the hills, forests, and streams of the Pacific coast. Fish grow fat in the ocean and transport those nutrients upstream in their bodies, depositing them inland. The salmon do not just swim in rivers, *they are rivers*—rivers of biomass running in the opposite direction from the water, rivers the Indigenous population tapped to support themselves. As long as those rivers flowed, the people of the Pacific Northwest would have everything they could ever need.

That's why they had to be dammed. European-descended colonizers arrived in force during the nineteenth century and immediately grasped that the Indigenous population's resilience lay in their prodigious fishing. If they were to be defeated and separated from their ancestral lands—as was always the first objective of American and Canadian colonizers, in every region—they would first have to be separated from the gift-bearing salmon. The very first fishing restrictions laid down by colonial governments in both the American Northwest and

British Columbia banned the deployment of fishing weirs on the grounds that the structures caused the depletion of salmon stock. Obviously false, this claim; Indigenous people had used them for thousands of years. Yet it served as an effective pretext to harm those populations while sparing white fishers, who did not build fishing weirs. Next the colonial governments passed laws forbidding Indigenous fishing for trade or sale. They could still fish for food, but this distinction went unrecognized by the tribes who relied on preserved salmon as their primary fungible good. Their entire way of life had effectively been outlawed.

White settlers kept up the pressure, requiring government permits even for sustenance fishing. They ceased to honor any distinction between Indigenous and colonial fishers, subjecting them all to the same restrictions with regard to closed seasons, closed areas, and banned gear. Gradually the healthy and wealthy nations of the Pacific Northwest diminished. European diseases claimed many lives, as elsewhere. Without full access to their fisheries, the tribes had no realistic chance to recover. Under severe economic duress many people abandoned their ancestors' lifeways for the poorly paid drudgery of labor in commercial salmon canneries owned by colonial elites. The same fish that gifted power and independence to their grandparents became an instrument of their oppression. As occurred with the Great Plains buffalo, Indigenous populations saw their long-established economies casually destroyed by white settlers who stole their natural resources and forced them to sell their hourly labor instead. The cannery business exploded. In 1875 fourteen canneries operated along the Lower Columbia, and five years later settlers had built another fifteen.[17] By the end of the 1800s every sober observer agreed that the salmon were overfished.

Zero Sum

In the earliest decades of the twentieth century, simple earthen dams began to appear on the Columbia River's upper tributaries, built to impound water for agriculture or to safeguard against flooding. An adult salmon driven by breeding rage can jump up to twelve feet vertically out of the water, which lets the fish clear logjams, waterfalls, and other obstructions they might meet on the journey to their spawning grounds.[18] They can clear weirs and small dams, but the Age of Big Dams that began in the 1930s exacted a brutal toll on the Northwest's salmon. World War II prompted a massive hydroelectric build-out to power aluminum smelters and aircraft factories (see chapter 2). The federal government poured hundreds of millions of dollars into damming nearly every major river in the region, carving these watery ecosystems into separate sections the salmon could not traverse. Adult fish swimming upstream could not clear the concrete barriers; juveniles spawned and headed downstream only to get pulped in hydroelectric turbines. Pacific salmon lost most of their ancestral territory in the span of a few decades.

What spawning grounds remained suffered from other consequences of dambuilding. Reservoirs store up frigid snowmelt, which keeps the rivers below the dams running shallow and warm, and salmon cannot spawn if the water is too warm. Warm water spreads disease among salmon and encourages algae to bloom, releasing toxins into rivers and depriving them of oxygen once the blooms die off. Dams lock up nutrients that should be seeding the environment around the salmon nests, leaving the new fry with less to eat once they hatch. Pollution kills adult fish and juveniles alike. The whole family of salmonids is vulnerable to human activity on account of their dependency on freshwater tributaries (what scientists call an anadromous lifestyle). They cannot avoid contact with human civilization. As a consequence many Western dams feature salmon hatcheries on the riverbank just below them, built by state governments to keep the runs on life support even as most returning fish die bewildered and helpless at the dams' feet.

Drought exacerbates every one of these problems by constricting resources to the point where humans and animals must compete directly for fresh water. Salmon will always lose these conflicts. As the twenty-first-century megadrought applies its python grip, irrigation districts from Washington to California face unenviable choices between satisfying local growers and pushing salmon runs toward extinction. The Bureau of Reclamation wants a voice in these discussions. So do local state and national politicians. So do environmental groups. So do the Indigenous tribes that still hold fishing rights on these rivers.

The Klamath River coalesces in southeastern Oregon at the base of the Cascade Mountains. Marshy meadows collect run-off each spring, which drains downhill through dozens of tributaries before landing in the Klamath's well-worn gorge. The river canyon cuts across high desert and into California, skirting a ridge above the Central Valley to the south. At this point the Klamath consumes the outflows of the Shasta and Trinity Rivers, swelling high, rushing with urgency down a widened channel to the Pacific Ocean south of Crescent City, California. It's the biggest and most important river south of the titanic Columbia and north of the mechanically augmented Sacramento. Several million people drink its water and so do the potato, alfalfa, and hay farmers who have converted most of the seasonal wetlands into agriculture. Some 75 percent of birds traversing the Pacific Flyway (the most important migratory network in the West) feed and rest in what remains of the Klamath's upper basin.[19] In a high and arid region, the Klamath is everyone's mother.

Of course that holds true for the salmon. Historically the Klamath hosted up to one million adult salmon and steelhead trout returning home each year, good enough for third place among western rivers (the Columbia has always held the top spot, historically returning up to *16 million* anadromous fish annually).[20] In the 2020s more than 90 percent of the Klamath's salmon are gone, largely as the consequence of dams. Six dams block the river's flow, the lowest of which is the aptly named Iron Gate Dam several miles east of highway I-5 and just south of the Oregon border. Salmon run unimpeded for hundreds of miles to that point, keeping the population extant, but their waters are warmer and less hospitable than at any time during their evolutionary history.

Iron Gate went up in 1964, at the height of Floyd Dominy's Ozymandian dam-hoisting campaign and over the furious objections of the Yurok Tribe. The Yurok count themselves the largest Indigenous tribe in California with more than six thousand members; their reservation comprises ancestral lands on both banks of the Klamath River from its mouth to the town of Weitchpec forty-four miles away. More than a thousand Yuroks live on the reservation, and, until 1964, they harvested enough salmon to feed themselves and make a decent commercial fishing income as well. Frankie Myers, a Yurok leader, told the Sierra Club, "Before the dams went up, we did have large commercial fisheries around the Klamath, but we didn't start to see a sharp decline in returning runs until 1964 when Iron Gate came in."[21] The dam is far upriver from the reservation, yet it dramatically impacted how many adult salmon made it to the ocean.

Today the median income of the Yurok reservation is about $11,000. Of reservation residents 35 percent live in poverty and 90 percent experience some degree of food uncertainty. The unemployment rate stands at 80 percent.[22] Parents who ate salmon every meal in their youth find they cannot catch or buy enough to feed their children. Obesity rates on the reservation skyrocketed, which is no paradox when one considers families accustomed to lean protein spending their limited food budgets on cheap sugars and starches. A community without access to food or income cannot survive. The law requires the private company that operates the dams to periodically release water downriver to nourish salmon runs, and generally they're happy to comply, but drought complicates the transaction. The Klamath River Indigenous peoples spent decades wrangling with the federal government in court over how much water they were entitled to keep in the river. In 2013 they fully asserted their water rights for the first time, which upset farmers who chafed seeing water go to fish (and by extension, Indigenous communities) rather than to irrigators. Farmers and ranchers have gone so far as to force open irrigation canals, which lasts as long as it takes for federal authorities to close them up again. Federal and state governments promised western landowners a certain level of prosperity for decades; those owners are now overleveraged with debt and struggle to accept that those promises were built on sand.

A Crack in the Wall

The combination of the Klamath dams and the megadrought has pressured the Yuroks like nothing before in their history. But in the last few years the Yuroks, and other Indigenous tribes that fish in the Klamath's tributaries, have kept their eyes on a ray of hope: the Klamath dams are coming down. Four of them anyway, the lowest four on the river: Iron Gate Dam, John C. Boyle Dam, and a pair of dams called Copco 1 and 2. The remaining two, smaller structures placed very high in the basin, have reasonable accommodations for migrating fish.

As dams across the United States reach the end of their planned lifespans, more than a few communities are opting to remove the structures and cut their rivers loose. However solid their construction might have been in the early to mid-twentieth century, decades of low public spending and steadily worsening storms have left them in need of reinvestment. One cannot just leave a dam to molder—it is a complex machine with many components under constant stress from very powerful forces. Neglect makes failure inevitable, whether that failure takes the form of an overtopping or a breach or anything else. Either the dam's hydropower and flood control is worth the millions of dollars that repairs and upgrades will cost, or it is not worthwhile and the dam must come down. In a political structure that consistently produces stasis, the end of a dam's life brings a refreshing kind of clarity. Sixty-nine American dams came down in 2020, most of them in the East where dams tend to be older, smaller, and less economically impactful.[23] Ninety thousand dams are on American soil, almost all of them pushing the end of their planned lifespans.

The Klamath River's dams are old too, some approaching a century. They do not produce much electricity anymore, especially with the river at such low ebb. They don't protect any population centers from flooding. Like most western dams, they were built explicitly for economic reasons. They do not provide drinking or irrigation water. For decades environmentalists and Indigenous groups urged the Department of the Interior to decommission the Klamath's dams. The salmon and other river life, they argued, hold more enduring value than four obsolete hydroelectric dams. In 2010 an impressively large cross-section of government officials gathered to announce and celebrate an historic decision: the Klamath dams would come down. The governors of Oregon and California spoke alongside state legislators, Indigenous leaders, and environmentalists.[24] All celebrated the historic ambitions of their shared project: four large dams on one river, opening more than four hundred miles of habitat to anadromous fish, restoring the ancient conveyer of nutrients between the Pacific Ocean and the Cascade Mountains. Planners project a total cost of $340 million—a large sum but less than the cost of upgrading all the dams, some of which date to the 1920s.[25] By 2024 the four condemned dams should be gone, though the COVID epidemic has already pushed the timetable back. Every year of drought drives the Klamath salmon populations closer to extinction.

In exchange the salmon will get their river back. The Klamath tribes will receive the opportunity to rebuild at least a fragment of their most important economic resource. Gradually dam operators will empty the reservoirs, releasing water only during winter months once the season's fry have spawned (lest its sediment smother their eggs). Three of the dams are concrete, so engineers will use drills and carefully placed explosives to fracture them into chunks for piecemeal removal. Iron Gate's locally sourced clay will return to the crater from which it was dug. Workers will seed the barren, scoured riverbeds immediately downstream of the dams with native plants that will anchor the newly free-flowing sediment into healthy riverbanks.[26] Every stakeholder, even the electric utilities, would love to see the Klamath become a model for river basin recovery in the West. For all the damage human beings can do to nature, we should take great comfort in nature's ability to recover. Time and again, from the Chernobyl exclusion zone to coral reefs in the Philippines, nature has demonstrated that, in the absence of damaging human activity, it can heal itself with astonishing speed. Along the Klamath River it may yet receive that opportunity. The salmon, silver as Pacific clouds and strong as the waves, the most generous of guests, have certainly earned it.

11

NO PROMISES

Near the end of each calendar year, November or so, I mount my neglected gray Fuji and ride it three miles to the American River Gorge. It is still warm so long as the sun shines, and occasionally one can spot bald eagles come to visit from the nearby Sierra Nevada. There is a paved and well-trafficked path along the river's north bank, thirty feet up a gorge wall made steeper by nineteenth-century hydraulic mining. I lock my bike to the guardrail and gingerly descend the gravel slope to the riverbank. I stand at the foot of the Nimbus Dam on a fringe of slippery stone to watch the salmon die. I watch because there may come a day in my lifetime when they stop running and never come again.

Chinook and steelhead, the latter actually a type of trout, run the American River. Both spawn in frigid freshwater tributaries, migrate to the Pacific for adulthood and then haul their ocean-fed bodies back upriver on terminal runs to their birthing grounds (see chapter 10). The fish I observe each November have crossed the Golden Gate of San Francisco twice and have swum up the Sacramento and then the American Rivers until running hard up against the Nimbus Dam. The state of California built the Nimbus Fish Hatchery in 1958, attempting to compensate for the loss of over one hundred miles of spawning grounds along both forks of the American River, all of which became impassable once the Bureau of Reclamation built the Nimbus Dam and its larger upstream counterpart, the Folsom Dam in 1955.[1] Hatchery workers capture the largest females they can find, squeeze out their brilliant orange roe and fertilize the eggs in artificial nests before releasing the young fish to their migrations.

The salmon I watch from the bank are these same fish, returning years later to a river that defies them. They want to breed in the same streams their ancestors did, walled off since the 1950s. Generation after generation spawns in the hatcheries with false memories of streams even their great-grandparents never actually saw, and they will kill themselves trying to get back there. They arrive dark and speckled, their fine-toothed mouths hanging open, gills red and angry. There are so many they run in clusters, knotting together fin-to-fin and teeming at the water's surface. Even from the bike path you can watch the river seethe with their bodies. Fearlessly driven onward, the salmon ignore my presence. They swim so close to the bank that I can dip my hand and feel the cold filaments of their dorsal fins. Men in yellow hip waders stand in the shallows of the opposite bank plucking hapless fish out of the still water at the dam's foot. It seems unsporting, but all these fish are doomed. They will never get where they are trying to go. They cannot understand that the maps coded inside their brains' olfactory centers no longer exist.

The 2020s are unfolding as a series of interlocking disasters, and it is hard to imagine that changing. The problems looming over the human race are so large that they can only be attacked over the course of years, by slow degrees, by the bending of second-order derivatives. These challenges would daunt any generation of people, at any time in history. To face them now, with so many governments around the world flailing and delegitimized from other crises, does not inspire confidence. In the United States we have almost totally dismantled our state's capacity to respond effectively to problems. Our government struggles to advance any serious objectives, and the U.S. Congress no longer functions except as a path to celebrity for its most venal members. In that vacuum, legislating falls to the president or an aging and increasingly extreme judiciary. The global COVID pandemic that began in the early months of 2020 revealed the impotence of the U.S. government better than could any written polemic. The People's Republic of China, where the disease first took root, used decisive measures to control COVID's spread and (like more than a few other nations such as Vietnam, Australia, and New Zealand) successfully limited deaths. More than one million Americans perished. Our political dysfunction is so pronounced that not only have numerous states effectively declared themselves in open defiance of federal public health policy—several have even dismantled their own protocols toward infectious disease. The state of Florida is currently attempting to remove vaccine requirements for diseases like the measles in public schools, due to a chain of culture-war logic I will not demean the reader by explaining.

The American people are not unlike the salmon I cycle out to see each November: we have run up hard against the limits of what our system of governance can provide. We have spent nearly fifty years deregulating business and slashing public investment while bankers and military contractors amassed tax-free fortunes. We have squandered enough resources to build a post-scarcity world twice over. Whatever mechanisms the Constitution offers have utterly failed (there will never be another Constitutional amendment ratified, for example), and our ostensibly democratic institutions have grown impervious to public opinion. Issues like marijuana legalization, which polls north of 70 percent nationwide for recreational use and 90 percent for medical use, tumble for generations in a culture war that has replaced any serious discussion of national policy. The civics lessons we learned in high school do us no more good than a salmon's ancestral memory of a breeding ground it cannot reach. The maps in our heads have nowhere left to guide us.

Everything Is Infrastructure

Hurricanes and floods in the East, drought and fire in the West, economic stagnation everywhere, and a state that's dismantled itself from the national to local level. In some circumstances, facing thorny crises on all sides, one might be tempted to set priorities and attack one at a time. With climate change that impulse is entirely wrong. When a structure threatens to collapse, it's not enough to shore up only the weakest strut. Because all of our problems intersect with one another, we must confront all of them simultaneously. Otherwise the problems we leave unfixed will undermine our labors elsewhere. The work of guaranteeing everyone a living wage and mitigating climate change must be the same work, or the effort will be wasted and the campaign for a better world will splinter and fail in the face of an opposition with no scruples and bottomless pockets. Either we bring everyone together into the future or no one at all.

So how does it begin? We'll start by acknowledging that the United States has always derived its prosperity from public investment. Infrastructure never came naturally to a frontier colony built along the British model of extracting maximum value from stolen land. Even Thomas Jefferson balked at the Erie Canal. The Mississippi Delta endured annual floods for centuries, yet none of the planters turned their riches to flood control because they thought of their land and people as assets to exploit. The cities of the Ohio Valley and the Great Lakes owe their existence to navigable waterways, none of which were reliable until the state stepped in to build canals and sweep rivers for snags. New York could not become a megacity until it built itself above the ocean and drained its wetlands as did New Orleans down south. The mountainous West would look totally different if not for expensive federal projects that built interstate highways and used dams to distribute irrigation water and electricity. Cheap electricity from Bureau of Reclamation dams built the modern Pacific Northwest. Las Vegas would not exist without the Hoover Dam. The Southwest's present population would be unsupportable without government water projects. Almost every region of our country has been rebuilt in the pursuit of collective prosperity. That is what it means to be a society.

The United States became a modern nation-state because we spent almost a century building infrastructure. Roughly fifty years ago that project ground to a halt; financial speculation and defense contracting rose to replace it. The results are plain to see. We must model our future on the widely distributed prosperity of the New Deal era: a policy arsenal so popular across this big diverse country that its disciples cemented themselves into power for decades. We cannot build the same things they did—the good dam sites are gone, rural states already electrified and so on—but even a casual glance around the United States reveals a nation begging to be rebuilt.

In the twenty-first century we must view infrastructure through a wide aperture. We must expand our definition of infrastructure beyond roads, bridges, and railways the same way our ancestors had to think bigger than canals and river snags. Anything people need that the free market will not provide is infrastructure. If you can't get your goods to market, a road or a canal is infrastructure. If you need more water to keep your ranch solvent than falls from the sky, an irrigation dam is infrastructure. If you suffer from health problems and cannot secure a job that offers benefits, then healthcare is infrastructure. If the water flowing from your tap is not safe to drink, then new pipes and water treatment plants are infrastructure. If you can't afford rent in the city where you work, then housing is infrastructure. If you can't reliably get online to access the many services governments and utilities have moved online, then public broadband and wireless networks are infrastructure. If your neighborhood is host to a now-toxic industrial waste site, then cleanup programs like Superfund are infrastructure. If heatwaves leave that same neighborhood's streets unwalkable in the summer, then new trees are infrastructure. Every one of those programs is a good idea on the merits that they would all improve people's lives, but, *even if you don't care about that*, it is undeniable that doing these things would enhance Americans' work output. They would all boost our productivity and economic capacity, if that's the highest priority. We do not do them because of exactly the crisis described in the last few chapters: a society crippling itself through disinvestment for the benefit of billionaires.

We need to get busy on these projects. There are lifetimes of work to be done and countless millions of good people eager for direction. Cost matters little, if at all; rarely in history has the U.S. government been able to print and borrow money so cheaply, and never has the need been more dire. For decades conservatives have used the specter of budget deficits and late-1970s inflation to undermine public investment, as working Americans increasingly disengage from politics. Give those people a reason to support a party program once again. Offer them real benefits, tangible improvements to their lives, and the resulting electoral boost could go a long way toward solving all these problems at once. An ambitious program that does not aggressively court the approval of otherwise apolitical citizens cannot expect success. Since the stakes are nothing less than the survival of American society, we cannot afford to undershoot our targets. There is no sum of money we can possibly spend that will exceed the damage climate change will do to our society. Every single dollar spent on decarbonizing our society or mitigating climate harm will create many dollars in benefits. As comedian Tim Heidecker famously declared, "it's free real estate."

Actionable Items

Planning out U.S. infrastructure and industrial policy for the next few decades seems like daunting work, but thankfully most of the intellectual lifting has already been done by left-leaning thinktanks and Congressional staffers. The "Green New Deal," as it has cleverly been pitched, is not a panacea for the nation's problems. Among its flaws: it includes no practical timetable for dismantling the domestic fossil fuel industry, an absolutely necessary task if the human race is to survive. But the Green New Deal, which is a very readable and relatively short piece of legislation that anyone can examine online, lays out a program for long-term investment in the working people of the United States.[2] It doesn't paint a complete picture, but it is a good answer for how to fill the gaps that decarbonizing will leave in our economy.

WATER SUPPLY: The West will continue to suffer from the multi-decade megadrought currently underway. As a consequence almost every population center west of the Mississippi is trying to secure additional water supplies. Without coordination and aggressive investment from the federal government it's easy to anticipate the zero-sum pile-up that will result, as wealthy cities deploy their financial leverage to outbid their poorer neighbors. Agriculture will likely need to take a step back in many regions, or at least reinvest in crops with smaller water profiles. Fewer almonds and pistachios and cattle, more corn and beans and potatoes. Historically important agreements like the Colorado River Compact will need rewriting in an environment where hot weather leads to dry soil which blocks snowpack from joining rivers. The Colorado's Upper Basin received 100 percent of its typical snowpack in 2020, yet only 50 percent of its typical runoff. In 2021 those numbers were 90 percent and 30 percent.[3] Dams, reservoirs and other surface projects won't yield much benefit; we have to collect and conserve what little water comes our way. Storm drains can be rebuilt as "bioswales," landscaped and vegetated strips of terrain that soak floods like sponges and help recharge groundwater. Tree-planting programs, rain gardens, permeable pavements and green or "living" roofs can all make a difference at the margins.

RETROFITTED BUILDINGS: Most of the commercial structures we work and shop in could be far more efficient in any number of ways. Urban wastewater treatment could reclaim significantly more water than it does if those departments had the resources to rebuild with new technology. Better plumbing and insulation, newer equipment, all-electric heating, improved HVAC systems, solar panels on rooftops and in parking lots— most merchants and developers would happily make these changes if handed money to do so. That work would provide an immediate lift to local economies and contractors even in parts of the United States that have suffered due to deindustrialization. Given the sheer number of buildings spread across this enormous country and how many are in need of investment, it's an immediately scalable idea that would make a visible impact on communities that federal policy has historically left behind.

FIRE AND FLOOD CONTROL: How we respond to these growing threats will vary from one region to another. Coastal cities will need to turn their shorelines into ramparts against sea level rise while smaller communities in Appalachia take down their dams, digging canals that feed protected flood plains. Northern California needs to move its entire electrical grid underground to limit the horrific fires that have begun to claw all the way across the Sierra Nevada. The Pacific Gas & Electric Company (PG&E), that ancient foe of Californians, has caused incalculable damage and ought to see its resources placed securely in the hands of the public. PG&E has argued that safely maintaining their vast network is practically impossible, and even if that is true it means the state is the only actor capable of shouldering such a grave responsibility. The states of the West badly need to upgrade their wilderness firefighting capacity; as of 2021 many of them, especially California, supplement their professional firefighting corps with teams of incarcerated people paid pennies on the dollar for labor that is grueling, specialized and dangerous all at once. When they finish their sentences, they are barred from working as professional firefighters. Government firefighting budgets must expand if this barbaric practice is to end.

TRANSPORTATION: Moving from a high-carbon society to a zero-carbon society will entail, among other things, overhauling our transit networks. The alterations many cities made to their downtown spaces during the COVID epidemic (shutting down certain streets to traffic, converting others to bicycle lanes and expanding restaurant footprints onto the street) were all positive steps on their own merits. Public transit needs major investments from the municipal level to the interstate level—it should not be cheaper and more convenient to travel by air than by rail in densely populated regions like the Northeast. If cheap and effective mass transit options exist, Americans can finally leave behind the stress and expense attached to their cars. Complementing this effort, cities must dismantle the awful wastelands of highway that cut them apart fifty or sixty years ago and rezone to resurrect the urban neighborhoods that gave the United States so many of its greatest citizens. "Linking higher density land use and transportation is one of the most effective tools available for cities to reduce their [carbon] footprint," declares the Association of Washington Cities.[4] Nor can public transit simply act as a conduit between the home and the workplace. People ought to be able to work and play at distances near and far, without isolating themselves for a drive in an automobile.

Cities and states would act on many of these items quickly if given the resources. In the United States, municipal and state-level governments always labor under fiscal restraints imposed by their medium-term revenues. Many are bound by law to balance their budgets each fiscal year, rain or shine. Yet the U.S. federal government, because it controls the world's preferred unit of exchange (American dollars) along with the sole entity empowered to control American dollars (the Federal Reserve), can effectively set the terms of its own borrowing. When the fed rescued the financial industry in 2008–9, and again during the COVID-driven financial panic of 2020, at a keystroke the money it "lent" simply appeared in the banks' accounts. Shifts in consumer demand and war overseas spiked the costs of key commodities in the years following, but for all the invocations of "inflation," the dreaded wage-price spiral failed to manifest and long-term inflation projections stayed low. The federal government faces no practical limit to its spending capacity, unlike the states and cities that make up the United States, so it must step in and bridge funding gaps wherever they appear. The National Infrastructure Advisory Council estimates that costs for water infrastructure alone will exceed available capital by up to $1 trillion.[5] It needs to be done, so the federal government will either pony up what is needed or watch the last physical legacy of the New Deal disintegrate in its hands.

In a wholly rational world this would be no choice at all. We would just do it. But then again, a society that made constructive long-term decisions would never have found themselves (ourselves, in this case) in such a bleak predicament. We defunded our own government on the grounds that it kept us from being rich and free, but it was always a long con by the descendants of the tycoons who opposed Franklin Roosevelt, and, now in the face of disaster, we are totally impotent. But we weren't always. The era of relatively competent and compassionate government, of a state that could chart and follow a course, still stands within living memory. Other countries govern themselves respectably well. It can be done. You do not need a nation of philosophers governed by saints.

How did we end up here, from there? Why have we failed, as a society, to progress? In the 2008 film *Iron Man*, a villainous

Jeff Bridges, frustrated by his scientists' attempts to re-engineer the miraculous power source propelling Iron Man's suit, bellows, "Tony Stark built this in a cave! *Outta scraps!*" Being an American citizen often feels this way. Our ancestors built incredible things under seemingly impossible conditions—works that shape our lives generations later—and yet today we cannot meaningfully address any serious problems.

Since the 1970s American government has divested in infrastructure, generally with the silent assent of a population taught to distrust anything calling itself *public* and undereducated in its own recent history. But that history is sorely needed. I chose dams as the subject for this book because of the vital role they played in making the New Deal and the United States successful. Our recovery from the Great Depression, our victory in World War II, and the decades of broadly shared prosperity that followed were the product of constant public reinvestment, of which ninety thousand dams represent only a sliver. That the age of dams came to an end just as the New Deal consensus collapsed into acrimony, inequality, and fear is no coincidence. The dams *were* the consensus, in ways both literal and metaphorical: a system of redistributing resources from the nation's wealthy Northeast to the hinterlands that held the rapidly expanding nation together with the promise of shared growth. We are not going to solve our present crises by building a bunch of dams, but we must recapture the spirit of our great infrastructure projects if our way of life is to endure. Almost everything we know in the twenty-first century, from our form of government to the landscapes of our cities, came about so recently it was practically yesterday. Anything about this world can change tomorrow if we really want it.

We are in this for the species, but in moments of pessimism I find great comfort in watching birds. No binoculars, no guidebook, just the house finches who teem at the feeders my wife keeps and the towhees who pick at hulls mere feet from the dogs. At the start of winter, magpies appear, and the mated pair of scrub jays who frequent our yard drive them off screaming. They don't care about a single thing that worries me, which lands as a sorely needed reminder that our planet is an overwhelmingly beautiful place. It is bigger and stronger than we are, and it will remain beautiful for billions of years. It does not depend on us. We do not have to save the world; that's just a phrase from the movies. Before us lies only the choice of whether or not to save ourselves, whether or not to build a society that protects and nurtures its people. This is not an all-or-nothing proposition—disaster is a gradient of infinite degrees, and every good thing we do is worthwhile. It will be tough work, and thankless, but one day the rewards may speak for themselves. The water is rising, the shovels are here. The life you save will be your own. No promises, all right?

NOTES

1. Rivers Wild

1. Billington, Jackson, Melosi, *The History of Large Federal Dams*, 1.
2. Shank, *Towpath to Tugboats*, 12.
3. Tarkov, "Engineering the Erie Canal."
4. McCullough, *Johnstown*, 75–77.
5. Coleman, Katkins, Wojno, "Dam-Breach Hydrology of the Johnstown Flood of 1889."
6. Conductor S. W. Keltz to Pennsylvania Railroad Investigators, Johnstown Area Heritage Association.
7. Conductor John Hess to Pennsylvania Railroad Investigators, Johnstown Area Heritage Association.
8. Shugerman, "The Floodgates of Strict Liability."

2. Brick by Brick

1. California Dept. of Parks and Recreation, "California's First Environmental Law."
2. AAA: Hiltzik, *Colossus*, 36.
3. Reisner, *Cadillac Desert*, 16–20.
4. Hiltzik, *Colossus*, 74.
5. Hiltzik, *Colossus*, 85.
6. Hiltzik, *Colossus*, 86.
7. Hiltzik, *Colossus*, 87.
8. Hiltzik, *Colossus*, 107.
9. Hiltzik, *Colossus*, 102.
10. Hiltzik, *Colossus*, 96.
11. Hiltzik, *Colossus*, 331.
12. Reisner, *Cadillac Desert*, 153–54.
13. Reisner, *Cadillac Desert*, 162.
14. Reisner, *Cadillac Desert*, 157.
15. Reisner, *Cadillac Desert*, 163.

3. The Drowned Empire

1. Barry, *Rising Tide*, 36.
2. Barry, *Rising Tide*, 45.
3. Barry, *Rising Tide*, 45.
4. Barry, *Rising Tide*, 48–49.
5. Barry, *Rising Tide*, 48.
6. Barry, *Rising Tide*, 49, 55.
7. Barry, *Rising Tide*, 53.
8. Barry, *Rising Tide*, 3.
9. Barry, *Rising Tide*, 16.
10. Barry, *Rising Tide*, 202.
11. Barry, *Rising Tide*, 201.
12. Barry, *Rising Tide*, 278.
13. Barry, *Rising Tide*, 270.
14. Barry, *Rising Tide*, 262.
15. Barry, *Rising Tide*, 274.
16. Barry, *Rising Tide*, 280
17. Barry, *Rising Tide*, 284.
18. Barry, *Rising Tide*, 312
19. Barry, *Rising Tide*, 316.
20. Barry, *Rising Tide*, 368.
21. Barry, *Rising Tide*, 387.
22. Roosevelt, Inaugural Address.
23. Perkins, *The Roosevelt I Knew*, 245.
24. Ezzell, "Wilson Dam and Reservoir."
25. Lebergott, *The American Economy*, 269.
26. Davis, *Where There Are Mountains*, 13.
27. Hiltzik, *The New Deal*, 78.
28. Virginia Commonwealth University, "Tennessee Valley Authority."
29. Tennessee Valley Authority, *Our History*.

4. A Towering Height

1. Watkins, *Righteous Pilgrim*, 13.
2. Watkins, *Righteous Pilgrim*, 51.
3. Ickes, *The Autobiography of a Curmudgeon*.
4. Watkins, *Righteous*, 268.
5. Watkins, *Righteous*, 279.
6. Hiltzik, *The New Deal*, 20.
7. Hiltzik, *The New Deal*, 16–17.
8. Hiltzik, *The New Deal*, 53.
9. Hiltzik, *The New Deal*, 112.
10. Hiltzik, *The New Deal*, 161.

11. Hiltzik, *The New Deal*, 422.

12. Watkins, *Righteous*, 450.

13. Watkins, *Righteous*, 450.

14. Watkins, *Righteous Pilgrim*, 562.

15. Watkins, *Righteous Pilgrim*, 564.

16. Watkins, *Righteous Pilgrim*, 568.

17. Preston, *Vanishing Landscapes*, 39–40.

18. Preston, *Vanishing Landscapes*, 57.

19. Watkins, *Righteous Pilgrim*, 571–72.

20. Watkins, *Righteous Pilgrim*, 574.

21. Reisner, *Cadillac Desert*, 111.

22. U.S. Bureau of Reclamation, "Bureau of Reclamation."

23. Reisner, *Cadillac Desert*, 115.

24. Watkins, *Righteous Pilgrim*, 268.

25. Watkins, *Righteous Pilgrim*, 792.

26. Watkins, *Righteous Pilgrim*, 821.

27. Watkins, *Righteous Pilgrim*, 821.

28. Watkins, *Righteous Pilgrim*, 819–22.

29. Watkins, *Righteous Pilgrim*, 823.

30. Watkins, *Righteous Pilgrim*, 828–32.

31. Watkins, *Righteous Pilgrim*, 833.

5. Approaching the Spillway

1. Arnold, "The Legacy of Dam Architect Floyd Dominy."

2. Nichols, Glen Canyon, 11.

3. Powell, *The Exploration of the Colorado River*, 120–21.

4. Nichols, *Glen Canyon*, 35.

5. Marston, "Floyd Dominy: An Encounter with the West's Undaunted Dam-Builder," *High Country News*, August 28, 2000, https://www.hcn.org/issues/184/5980.

6. McPhee, *Encounters with the Archdruid*, 169.

7. Marston, "Floyd Dominy."

8. Marston, "Floyd Dominy."

9. Marston, "Floyd Dominy."

10. Marston, "Floyd Dominy."

11. Marston, "Floyd Dominy."

12. Foster, "Ecogeomorphology of the Grand Canyon."

13. U.S. National Park Service, "2019 A Good Year."

14. Henry Brean, "Researchers Study True Scale of Evaporation at Lakes Mead, Powell." *Las Vegas Review-Journal*, August 17, 2019, https://bouldercityreview.com/news/researchers-study-true-scale-of-evaporation-at-lakes-mead-powell-51822/.

15. California Department of Agriculture, *California Agricultural Statistics Review 2018–2019*, 2.

16. Reisner, *Cadillac Desert*, 340.

17. Reisner, *Cadillac Desert*, 115.

6. Great, Still Mass

1. U.S. Dept. of State, "Recognition of the Soviet Union, 1933."

2. Reisner, *Cadillac Desert*, 257.

3. Murray, "Smog Deaths."

4. Stewart Udall oral history interview, John F. Kennedy Presidential Library, March 16, 1970, https://www.jfklibrary.org/asset-viewer/archives/jfkoh/Udall %2c%20stewart%20l/jfkoh-slu-02/jfkoh-slu-02.

5. Cabell Phillips, "Cannon vs. Hayden: A Clash of Elderly Power Personalities in Congress," *New York Times*, June 25, 1962, https://timesmachine.nytimes .com/timesmachine/1962/06/25/82047246.html?pageNumber=17.

6. Phillips, "Cannon vs. Hayden."

7. U.S. Census Bureau, *Arizona Census*.

8. Perlstein, *Before the Storm*, 26.

9. Public statement by Dr. King on July 16, 1964, https://kinginstitute.stanford .edu/encyclopedia/goldwater-barry-m.

10. Reisner, *Cadillac Desert*, 274–75.

11. U.S. National Park Service official tally, https://grandcanyon.com/news /how-many-visitors-to-grand-canyon/.

12. Reisner, *Cadillac Desert*, 258.

13. Reisner, *Cadillac Desert*, 288–89.

14. Stewart Udall oral history interview, John F. Kennedy Presidential Library, March 16, 1970.

15. Maryland Dept. of Natural Resources, "Maryland's Lakes and Reservoirs."

16. Donald Janson, "Tocks Dam: Story of 13-Year Failure," *New York Times*, August 4, 1975, https://www.nytimes.com/1975/08/04/archives/tocks-dam -story-of-13year-failure.html.

17. Janson, "Tocks Dam."

18. Steve Novak, "Delaware River Loses Staunch Defender," *Lehigh Valley Live*, March 28, 2021, https://www.lehighvalleylive.com/warren-county/2021 /03/delaware-river-loses-staunch-defender-nancy-shukaitis-fought-the-tocks -island-dam-and-won.html.

19. Ryan Balton, "Controversy on the Delaware: A Look Upstream at the Tocks Island Dam Project," July 16, 2019, https://www.youtube.com/watch?v =xknxlDqxquU.

20. Meir Rinde, "Richard Nixon and the Rise of Environmentalism," *Science History*, June 2, 2017, https://www.sciencehistory.org/distillations/richard -nixon-and-the-rise-of-american-environmentalism.

21. Janson, "Tocks Dam."

22. Ed Hanley, "Behind the Scenes at the Creation of EPA," Environmental Protection Agency Alumni Oral History Program, https://www.epaalumni.org /userdata/pdf/600a1db1b9ef1e85.pdf.

23. Scott Shane, "For Ford, Pardon Decision Was Always Clear-Cut," *New York Times*, December 29, 2006, https://www.nytimes.com/2006/12/29/washington /29pardon.html.

7. The Breakdown

1. Reisner, *Cadillac Desert*, 389.

2. Reisner, *Cadillac Desert*, 383.

3. Reisner, *Cadillac Desert*, 397.

4. Reisner, *Cadillac Desert*, 397.

5. Reisner, *Cadillac Desert*, 399

6. Niagara Parks Service, "Niagara Falls Geology Facts & Figures" (site discontinued).

7. Reisner, *Cadillac Desert*, 404.

8. Reisner, *Cadillac Desert*, 406.

9. Grace Lichtenstein, "Residents are Divided on Rebuilding of Teton Dam," *New York Times*, January 2, 1977, https://www.nytimes.com/1977/01/02/archives /residents-are-divided-on-rebuilding-of-teton-dam.html.

10. Perlstein, *Reaganland*, 206.

11. Freeman, *The Wayward Welfare State*, 4–6.

12. NBC News, "Jimmy Carter Battles Plans for Dams—Again," July 28, 2008, https://www.nbcnews.com/id/wbna25889670.

13. Perlstein, *Reaganland*, 50–54.

14. Perlstein, *Reaganland*, 60.

15. Reisner, *Cadillac Desert*, 314–15.

16. Binyamin Applebaum and Robert Hershey Jr., "Paul A. Volcker, Fed Chairman Who Wages War on Inflation, Is Dead at 92," *New York Times*, December 9, 2019, https://www.nytimes.com/2019/12/09/business/paul-a-volcker-dead.html.

17. Applebaum and Hershey, "Paul A. Volcker."

18. Perlstein, *Reaganland*, 633.

19. Perlstein, *Reaganland*, 633.

20. Perlstein, *The Invisible Bridge*, 368.

21. Perlstein, *Reaganland*, 913.

22. Greider, "How James Watt Survives."

23. William Greider, "How James Watt Survives," *Rolling Stone*, June 9, 1983, https://www.rollingstone.com/politics/politics-news/how-james-watt-survives -68005/.

24. Dale Russakoff, "Watt's Off-the-Cuff Remark Sparks Storm of Criticism," *Washington Post*, September 22, 1983.

25. Julia Prodis, "James Watt Is Back in the Saddle," *Los Angeles Times*, May 5, 1991, https://www.latimes.com/archives/la-xpm-1991-05-05-me-1873-story.html.

26. Philip Shabecoff, "House Charges Head of E.P.A. with Contempt," *New York Times*, December 17, 1982, https://www.nytimes.com/1982/12/17/us/house-charges-head-of-epa-with-contempt.html.

27. Philip Shabecoff, "E.P.A. Chief Plans White House Legal Talks," *New York Yimes*, March 7, 1983, https://www.nytimes.com/1983/03/07/us/epa-chief-plans-white-house-legal-talks.html.

28. Dylan Matthews, "Paul Volcker, the Most Important Fed Chair of the 20th Century, Has Died," *Vox*, December 9, 2019, https://www.vox.com/policy-and-politics/2019/12/9/21002682/paul-volcker-dies-federal-reserve-volcker-rule.

29. "U.S. Military Spending/Defense Budget 1960–2022," *Macrotrends*, https://www.macrotrends.net/countries/usa/united-states/military-spending-defense-budget.

8. The War Economy

1. Cooper and Block, *Disaster*, 50–67.

2. Cooper and Block, *Disaster*, 72.

3. Cooper and Block, *Disaster*, 76.

4. Cooper and Block, *Disaster*, 80.

5. Cooper and Block, *Disaster*, 102.

6. Cooper and Block, *Disaster*, 23–26.

7. Cooper and Block, *Disaster*, 41–42.

8. Cooper and Block, *Disaster*, 44.

9. Interviews from Spike Lee's "When the Levees Broke," 2006.

10. Cooper and Block, *Disaster*, 132.

11. Cooper and Block, *Disaster*, 149.

12. Cooper and Block, *Disaster*, 147.

13. Cooper and Block, *Disaster*, 148.

14. Cooper and Block, *Disaster*, 211.

15. Cooper and Block, *Disaster*, 215.

16. CNN television broadcast, September 2, 2005, https://www.youtube.com/watch?v=x0tfv70v1uQ.

17. Cooper and Block, *Disaster*, 222.

18. Eric Lipton and Scott Shane, "Homeland Security Chief Defends Federal Response," *New York Times*, September 4, 2005, https://www.nytimes.com/2005/09/04/us/nationalspecial/homeland-security-chief-defends-federal-response.html.

19. Thomas Frank, "After a $14-Billion Upgrade, New Orleans' Levees are Sinking," *Scientific American*, April 11, 2019, https://www.scientificamerican.com/article/after-a-14-billion-upgrade-new-orleans-levees-are-sinking/.

9. Inundation

1. Stephen Labaton, "Congress Passes Wide-Ranging Bill Easing Bank Laws," *New York Times*, November 5, 1999, https://www.nytimes.com/1999/11/05/business/congress-passes-wide-ranging-bill-easing-bank-laws.html.

2. NBC News, "Number of Foreclosures Soared in 2007," January 29, 2008, https://www.nbcnews.com/id/wbna22893703.

3. Tom Dickinson, "No We Can't," *Rolling Stone*, February 2, 2010, https://www.rollingstone.com/politics/politics-news/no-we-cant-198655/.

4. David Dayen, "Obama Failed to Mitigate America's Foreclosure Crisis," *Atlantic*, December 14, 2016, https://www.theatlantic.com/politics/archive/2016/12/obamas-failure-to-mitigate-americas-foreclosure-crisis/510485/.

5. Frank Lysy, "Already Low Public Investment Has Fallen Sharply Under Obama," *An Economics Sense*, November 16, 2014, https://aneconomicsense.org/2014/11/16/already-low-public-investment-has-fallen-sharply-under-obama/.

6. Thomas Blanchet, Emmanuel Saez, and Gabriel Zucman, "Real-Time Inequality" (draft), January 30, 2022, https://gabriel-zucman.eu/files/bsz2022.pdf.

7. National Air and Space Administration, "Hurricane Sandy (Atlantic Ocean)," March 7, 2013, https://www.nasa.gov/mission_pages/hurricanes/archives/2012/h2012_Sandy.html

8. Miles, *Superstorm*, 210.

9. Miles, *Superstorm*, 245.

10. Miles, *Superstorm*, 210.

11. Lixion Avila and John Cangialosi, "Tropical Cyclone Report: Hurricane Irene," *National Oceanic and Atmospheric Administration*, December 14, 2011, https://www.nhc.noaa.gov/data/tcr/al092011_Irene.pdf.

12. Miles, *Superstorm*, 281.

13. U.S. Geological Survey, "Hurricane Sandy—New York," December 1, 2017, https://www.usgs.gov/centers/ny-water/science/hurricane-sandy-new-york?qt-science_center_objects=0#qt-science_center_objects.

14. Miles, *Superstorm*, 324.

15. Mireya Navarro, "New York is Lagging as Seas and Risks Rise, Critics Warn," *New York Times*, September 10, 2012, https://www.nytimes.com/2012/09/11/nyregion/new-york-faces-rising-seas-and-slow-city-action.html.

16. U.S. Geological Survey, "Hurricane Sandy—New York." National Hurricane Center.

17. Patrick McGeehan and Griff Palmer, "Displaced by Hurricane Sandy and Living in Limbo," *New York Times*, December 6, 2013, https://www.nytimes.com/2013/12/07/nyregion/displaced-by-hurricane-sandy-and-living-in-limbo-instead-of-at-home.html

18. Karina Hernandez, "Hey Moment for NYC's Post-Sandy Flood Protection Plans," *City Limits*, February 19, 2019, https://citylimits.org/2019/02/19/key-moment-for-nycs-post-sandy-flood-protection-plans/.

19. "Live Coverage of Hurricane Ida," *New York Times*, September 2021, https://www.nytimes.com/live/2021/09/02/nyregion/nyc-storm.

20. Jesse McKinley, Dana Rubinstein, and Jeffery C. Mays, "The Storm Warnings Were Dire; Why Couldn't New York Be Protected?," *New York Times*, September 3, 2021, https://www.nytimes.com/2021/09/03/nyregion/nyc-ida.html

21. CNN, "New York City Was Never Built to Withstand a Deluge Like the One Ida Delivered," September 5, 2021, https://wtop.com/weather-news/2021/09/new-york-citys-illegally-converted-apartments-proved-deadly-in-idas-path/.

22. U.S. National Park Service, "Flood Control by Wetlands."

23. U.S. National Park Service, "History of Jamaica Bay."

24. Lisa M. Collins, "Why N.Y.C. Is Humming with Wildlife," *New York Times*, October 28, 2021, https://www.nytimes.com/2021/10/28/nyregion/nyc-wildlife-habitats-animals.html.

25. Aaron D. Backes, "History of New York's John F. Kennedy International Airport," *Classic New York History*, 2020, https://classicnewyorkhistory.com/history-of-new-yorks-john-f-kennedy-international-airport/.

26. Andrew Freedman, "U.S. Airports Face Increasing Threat From Rising Seas," *Climate Central*, June 18, 2013, https://www.climatecentral.org/news/coastal-us-airports-face-increasing-threat-from-sea-level-rise-16126.

27. Malcolm J. Bowman, "A City at Sea," *New York Times*, September 25, 2005, https://www.nytimes.com/2005/09/25/opinion/nyregionopinions/a-city-at-sea.html.

28. Brahmjot Kaur, "New York Embarks on a Massive Climate Resiliency Project to Protect Manhattan's Lower East Side from Sea Level Rise," *Inside Climate News*, June 25, 2021, https://insideclimatenews.org/news/25062021/new-york-manhattan-coastal-resiliency-sea-level-rise/.

29. Collins, "Why N.Y.C. Is Humming with Wildlife."

30. Alyssa Battistoni, "The Flood Next Time," *Jacobin*, December 21, 2012, https://www.jacobinmag.com/2012/12/the-flood-next-time

31. Christopher Flavelle, "As Manchin Blocks Climate Plan, His State Can't Hold Back Floods," *New York Times*, October 17, 2021, https://www.nytimes.com/2021/10/17/climate/manchin-west-virginia-flooding.html.

10. Desiccation

1. Jeff Berardelli, "Western U.S. May Be Entering Its Most Severe Drought in Modern History," *CBS News*, April 12, 2021, https://www.cbsnews.com/news/drought-western-united-states-modern-history/.

2. Henry Fountain, "In a First, U.S. Declares Shortage on Colorado River, Forcing Water Cuts," *New York Times*, August 16, 2021, https://www.nytimes.com/2021/08/16/climate/colorado-river-water-cuts.html.

3. U.S. Bureau of Reclamation, "Lake Mead at Hoover Dam."

4. David Charns, "Body Found in Barrel in Lake Mead May Date Back to 1980s, More Likely to Appear as Water Recedes, Las Vegas Police Say," KLAS-TV Las Vegas, May 2, 2022, https://www.8newsnow.com/i-team/i-team-body-found-in-barrel-in-lake-mead-may-date-back-to-1980s-more-likely-to-appear-as-water-recedes-las-vegas-police-say/.

5. Brandon Loomis, "The Sand Is There, But Low Water Levels Halt a Controlled Flood to Restore Grand Canyon's Beaches," *Arizona Central*, November 1, 2021, https://www.azcentral.com/story/news/local/arizona-environment/2021/11/01/feds-say-theres-too-little-water-spare-grand-canyons-beaches/6201263001/.

6. Brian Maffly, "Worsening Drought Could Lead to Water-Use Restrictions," *Salt Lake Tribune*, May 24, 2021, https://www.sltrib.com/news/environment/2021/05/24/worsening-drought-could/

7. Reisner, *Cadillac Desert*, 534.

8. Reisner, *Cadillac Desert*, 530.

9. Reisner, *Cadillac Desert*, 474–75

10. Arizona Department of Environmental Quality, https://new.azwater.gov/conservation/agriculture.

11. Perez-Blancho et al., "Agricultural Water Saving," 1.

12. Perez-Blancho et al., "Agricultural Water Saving," 6.

13. Abrahm Lustgarten, "When the World Is on the Brink, $3.5 Trillion Is a Pittance," *New York Times*, October 28, 2021, https://www.nytimes.com/2021/10/28/opinion/climate-change-biden-spending.html.

14. Columbia River Inter-Tribal Fish Commission, "Salmon Culture of the Pacific Northwest Tribes," https://www.critfc.org/salmon-culture/tribal-salmon-culture/.

15. Columbia River Inter-Tribal Fish Commission, "Salmon Culture."

16. *Indigenous Foundations*, "Aboriginal Fisheries in British Columbia," http://indigenousfoundations.arts.ubc.ca/aboriginal_fisheries_in_british_columbia/.

17. *Oregon History Project*, "Salmon Decline, 1894," https://www.oregonhistoryproject.org/articles/historical-records/salmon-decline-1894/.

18. Beach, "Fisheries Research Technical Report," appendix 2.

19. *American Rivers*, "Klamath River," https://www.americanrivers.org/river/klamath-river/.

20. Hamilton et al., "Distribution of Anadromous Fishes in the Upper Klamath River Watershed," 10.

21. Joe Purtell, "Another Hurdle Cleared, Klamath Dams Closer to Coming Down," *Sierra Club*, January 14, 2021, https://www.theguardian.com/global -development/2021/oct/04/salmon-klamath-river-yurok-women-nutrition -health.

22. Lucy Sherriff, "'No Fish Means No Food': How Yurok Women Are Fighting for Their Tribe's Nutrional Health," *The Guardian*, October 20, 2021, https:// www.americanrivers.org/2021/02/69-dams-removed-in-2020/.

23. Jessie Thomas-Blate, "69 Dams Removed in 2020," *American Rivers*, February 18, 2021, https://www.americanrivers.org/2021/02/69-dams-removed -in-2020/.

24. Oregonian Editorial Board, "For Klamath Dams, It's Hasta La Vista," *Oregonian*, February 18, 2010, https://www.oregonlive.com/opinion/2010/02 /for_klamath_dams_its_hasta_la.html.

25. *American Rivers*, "American Rivers Applauds Oregon and California, Tribes, Pacificorp For River Restoration Leadership," November 17, 2020, https://www .americanrivers.org/conservation-resource/american-rivers-applauds-oregon -and-california-tribes-pacificorp-for-river-restoration-leadership/; Amy Souers Kober, "Could This Be the World's Biggest Dam Removal?" *American Rivers*, April 5, 2016, https://www.americanrivers.org/2016/04/worlds-biggest-dam -removal/.

26. Brian Oaster, "Will Klamath Salmon Outlast the Dam Removal Process?" *High Country News*, August 17, 2021, https://www.hcn.org/issues/53.9/indigenous -affairs-dams-will-klamath-salmon-outlast-the-dam-removal-process.

11. No Promises

1. California Department of Fish and Wildlife, "History of Nimbus Hatchery."

2. House Resolution 109, 116th U.S. Congress, https://www.congress.gov/bill /116th-congress/house-resolution/109/text.

3. Alex Hager, "These Four Metrics Are Used to Track Drought, and They Paint a Bleak Picture," KUNC Radio Northern Colorado, November 24, 2021, https://www.kunc.org/environment/2021-11-24/these-four-metrics-are-used -to-track-drought-and-they-paint-a-bleak-picture.

4. Association of Washington Cities, "Climate Resilience Handbook," 2021, 11.

5. Association of Washington Cities, "Climate Resilience Handbook," 2021, 14.

BIBLIOGRAPHY

Arnold, Elizabeth. "The Legacy of Dam Architect Floyd Dominy." *National Public Radio*, May 4, 2010. https://www.npr.org/templates/story/story.php ?storyId=126511368.

Barry, John M. *Rising Tide: The Great Mississippi Flood of 1927 and How It Changed America*. New York: Simon and Schuster, 1997.

Beach, M. H. "Fisheries Research Technical Report, No. 78." United Kingdom Ministry of Agriculture, Fisheries and Food, 1984.

Billington, David, Donald Jackson, and Martin Melosi. *The History of Large Federal Dams: Planning Design, and Construction in the Era of Big Dams*. Denver: Bureau of Reclamation, 2005.

California Department of Agriculture. *California Agricultural Statistics Review 2018–2019.* https://www.cdfa.ca.gov/Statistics/pdfs/2020_Ag_Stats_Review .pdf.

California Department of Fish and Wildlife. "History of Nimbus Hatchery." https://wildlife.ca.gov/Fishing/Hatcheries/Nimbus/History.

California {~?~IQ: Editorial: Selected paragraph may have unmatched curly quotation marks.}Department of Parks and Recreation. California's First Environmental Law," http://www.150.parks.ca.gov/?page_id=27596.

———. As cited in *Journal of Electricity* 1, no. 3 (September 1895). https://www .parks.ca.gov/?page_id=22909.

Carter, Jimmy. State of the Union Address to Congress, January 19, 1978. Transcript at: https://www.presidency.ucsb.edu/documents/the-state-the-union -address-delivered-before-joint-session-the-congress-1.

Clinton, William. Address to the 1992 Democratic National Convention. Transcript at: https://www.presidency.ucsb.edu/documents/address-accepting -the-presidential-nomination-the-democratic-national-convention-new -york.

Coleman, Neil, Uldis Kaktins, and Stephanie Wojno. "Dam-Breach Hydrology of the Johnstown Flood of 1889—Challenging the Findings of the 1891 Investigation Report." *Heliyon* 2, no. 6 (June 2016). https://www.ncbi.nlm.nih .gov/pmc/articles/PMC4946313/.

Cooper, Cristopher, and Robert Block. *Disaster: Hurricane Katrina and the Failure of Homeland Security*. New York: Henry Holt, 2006.

Crampton, C. Gregory. *Ghosts of Glen Canyon: History Beneath Lake Powell*. Salt Lake City: Bonneville, 2009.

Crandall, Warren Daniel. *History of the Ram Fleet and the Mississippi Brigade in the War for the Union on the Mississippi and Its Tributaries*. St. Louis: Buschart Brothers, 1907.

Davis, Donald Edward. *Where There Are Mountains: An Environmental History of Southern Appalachia*. Athens GA: University of Georgia Press, 2000.

Dimitry, Carolyn, Anne Effland, and Neilson Conklin. "The 20th Century Transformation of U.S. Agriculture and Farm Policy." *U.S. Department of Agriculture Economic Information Bulletin* 3 (June 2005). https://www.ers.usda.gov/webdocs/publications/44197/13566_eib3_1_.pdf.

Ezzell, Patricia Bernard. "Wilson Dam and Reservoir." In *Encyclopedia of Alabama*. http://www.encyclopediaofalabama.org/article/h-3268.

Foster, Lauren. "Ecogeomorphology of the Grand Canyon." UC Davis Center for Watershed Sciences. https://watershed.ucdavis.edu/education/classes/ecogeomorphology-grand-canyon-2016/.

Freeman, Roger. *The Wayward Welfare State*. Palo Alto CA: Hoover, 1982.

Greider, William. "Annals of Finance." *New Yorker*, November 9, 1987.

Hamilton, John, Gary Curtis, Scott Snedaker, and David White. "Distribution of Anadromous Fishes in the Upper Klamath River Watershed Prior to Hydropower Dams." *Fisheries* 30 no. 4 (2005). https://fisheries.org/docs/fisheries_magazine_archive/fisheries_3004.pdf.

Heidecker, Tim, and Eric Wareheim, "Free House for You, Jim." *Tim and Eric Awesome Show: Great Job!*, Season 4, episode 7, original air date March 24, 2009.

Hiltzik, Michael. *Colossus: Hoover Dam and the Making of the American Century*. New York: Simon and Schuster, 2010.

——— . *The New Deal: A Modern History*. New York: Simon and Schuster, 2011.

Ickes, Harold L. *The Autobiography of a Curmudgeon*. New York: Quadrangle, 1969.

——— . Introduction." In *Muddy Waters: The Army Engineers and the Nation's Rivers*, edited by Arthur Maass. Cambridge MA: Harvard University Press, 1951.

Johnstown Area Heritage Association Archives, Johnstown, Pennsylvania. Pennsylvania Railroad Interview Transcripts. http://www.jaha.org/edu/flood/story/prr_reports_flood.html.

Lebergott, Stanley. *The American Economy: Income, Wealth and Want*. Princeton NJ: Princeton University Press, 1976.

Maryland Dept. of Natural Resources. "Maryland's Lakes and Reservoirs: FAQ." http://www.mgs.md.gov/geology/maryland_lakes_and_reservoirs.html.

McCullough, David. *The Johnstown Flood*. New York: Simon and Schuster, 1968.

McPhee, John. *Encounters with the Archdruid*. New York: Farrar, Straus and Giroux, 1971.

Miles, Kathryn. *Superstorm: Nine Days Inside Hurricane Sandy*. New York: Dutton, 2014.

Murray, Ann. "Smog Deaths in 1948 Led to Clean Air Laws." National Public Radio, 2009.

Nichols, Tad. *Glen Canyon: Images of a Lost World*. Santa Fe: Museum of New Mexico Press, 1999.

Perez-Blanco, Dionisio, Adam Loch, Frank Ward, Chris Perry, and David Adamson. "Agricultural Water Saving through Technologies: A Zombie Idea." *Environmental Research Letters*, October 2021. https://iopscience.iop.org/article/10.1088/1748-9326/ac2fe0/pdf

Perkins, Frances. *The Roosevelt I Knew*. New York: Penguin Random House, 1946.

Perlstein, Rick. *Before the Storm: Barry Goldwater and the Unmaking of the American Consensus*. New York: Farrar, Straus and Giroux, 2009.

———. *The Invisible Bridge: The Fall of Nixon and the Rise of Reagan*. New York: Simon and Schuster, 2014.

———. *Nixonland: The Rise of a President and the Fracturing of America*. New York: Simon and Schuster, 2008.

———. *Reaganland: America's Right Turn, 1976–1980*. New York: Simon and Schuster, 2020.

Powell, John Wesley. *The Exploration of the Colorado River and Its Canyons*. New York: Simon & Brown, 1870.Preston, William L. *Vanishing Landscapes: Land and Life in the Tulare Lake Basin*. Berkeley: University of California Press, 1981.

Quarstein, John V. "Battle of Memphis." Mariners' Museum, May 20, 2021. https://www.marinersmuseum.org/2021/05/battle-of-memphis/.

Rattner, Steve. "Volcker Asserts U.S. Must Trim Living Standard." *New York Times*, October 18, 1979.

Reisner, Marc. *Cadillac Desert: The American West and its Disappearing Water*. New York: Penguin, 1986.

Roosevelt, Franklin D. Inaugural Address, March 4, 1933.

———. Address to U.S. Congress by President Franklin Roosevelt, April 10, 1933. Transcript: https://www.presidency.ucsb.edu/documents/message-congress-suggesting-the-tennessee-valley-authority.

———. Informal Extemporaneous Remarks at Montgomery, Alabama, January 21, 1933. Transcript: https://www.presidency.ucsb.edu/documents/informal-extemporaneous-remarks-montgomery-alabama-muscle-shoals-inspection-trip

Roosevelt, Theodore. Address to the U.S. Congress, January 12, 1907. Transcript available at https://www.presidency.ucsb.edu/documents/message-congress-the-threatened-destruction-the-overflow-the-colorado-river-the-sink-or.

Salon.com. Transcript of Q&A session at West Point, New York, August 6, 1974. https://www.salon.com/2015/10/14/libertarian_superstar_ayn_rand _defended_genocide_of_savage_native_americans/.

Shank, William. *Towpath to Tugboats: A History of American Canal Engineering.* York PA: American Canal and Transportation Center, 1982.

Shugerman, Jed. "The Floodgates of Strict Liability: Bursting Reservoirs and the Adoption of Fletcher v. Rylands in the Gilded Age," *Yale Law Journal,* November 2000.

Smith, Bessie. "Homeless Blues." Recorded 1927. http://www.protestsonglyrics .net/Historic_Songs/Homeless-Blues-Bessie-Smith.phtml. Audio here: https://www.youtube.com/watch?v=_VvGGSX3YtA.

Smith, J. P. "Louisiana Girding to Spare Project From the 'Hit List.'" *Washington Post,* March 28, 1977.

Tarkov, John. "Engineering the Erie Canal." *Invention and Technology* 2, no. 1 (Summer 1986). https://www.inventionandtech.com/content/engineering -erie-canal-2.

Tennessee Valley Authority. *Our History: The 1940s.* https://www.tva.com/about -tva/our-history/the-1940s.

U.S. Bureau of Reclamation. "The Bureau of Reclamation: A Very Brief History." https://www.usbr.gov/history/borhist.html.

——— . "Lake Mead at Hoover Dam, End of Month Elevation." https://www .usbr.gov/lc/region/g4000/hourly/mead-elv.html.

U.S. Census Bureau. *Arizona Census of Population, 1960s Decennial.* https://www2 .census.gov/library/publications/decennial/1960/population-volume-1 /vol-01-04-c.pdf.

——— . Grays Harbor County Report. 2021. https://www.census.gov/quickfacts /graysharborcountywashington.

——— . "The Great Migration, 1910 to 1970." September 13, 2012. https://www .census.gov/dataviz/visualizations/020/.

U.S. Department of State. "Recognition of the Soviet Union, 1933." https:// history.state.gov/milestones/1921-1936/ussr.

U.S. National Park Service. 2019. "2019 A Good Year for the Glen Canyon National Recreation Area." https://www.lakepowelllife.com/2019-a-good-year-for -the-glen-canyon-national-recreation-area/.

——— . n.d. "Flood Control by Wetlands." https://www.nps.gov/keaq/learn /education/flood-control-by-wetlands.htm.

——— . n.d. "History of Jamaica Bay." https://www.nps.gov/parkhistory/online _books/gate/jamaica_bay_hrs.pdf.

Van Green, Ted. "Americans Overwhelmingly Say Marijuana Should Be Legal for Medical Use." Pew Research Center, April 16, 2021. https://www.pewresearch .org/fact-tank/2021/04/16/americans-overwhelmingly-say-marijuana -should-be-legal-for-recreational-or-medical-use/.

Virginia Commonwealth University. "The Tennessee Valley Authority: Electricity for All" Social Welfare History Project. https://socialwelfare.library.vcu.edu/eras/great-depression/tennessee-valley-authority-electricity/.

Wallsten, Peter. "Chief of Staff Draws Fire from Left as Obama Falters." *Wall Street Journal*, January 26, 2010. https://www.wsj.com/articles/SB10001424052748703808904575025030384695158

Watkins, T. H. *Righteous Pilgrim: The Life and Times of Harold L. Ickes, 1874–1952.* New York: Henry Holt, 1990.

White, William S. *Majesty and Mischief: A Mixed Tribute to FDR.* New York: McGraw-Hill, 1961.

INDEX

Index